EARLY

by Mary Jane Nordgren, D. O.

```
  /\     /\      /\     /\         /\
 /  \   /  \    /  \   /  \       /  \
 //\\  //\\   //\\  //\\      //\\
 ) (    ) (     ) (    ) (         ) (
```

Cover Art by William A. Helwig

Copyright © 2000

EARLY

"Logging tales too human to be fiction"

Cover photos, clockwise:
Alfred Nordgren with workhorses Pat and Mike; Alfred in front of huge felled tree; Alfred with Earl and his dog Tygue (who was so protective of Earl, he wouldn't even let Hildur spank him) - early 1920's on the Home Place on Carpenter Creek; Earl and Al - late 1950's; Alfred in 1919 hard tired truck on wood plank road; Earl as high climber; Earl hanging from tongs

Photo facing:
The Nordgren Home Place
on Carpenter Creek in the 1920s

ISBN #0-9703896-0-4

Copyright 2000
by Mary Jane Nordgren, D.O.

Printed in the U.S.A. by Morris Publishing
3212 E. Hwy 30
Kearney NE 68847
800-650-7888

To Order, send $14.95 + $2.95 Postage to

TAWK Press
47777 Ihrig Road
Forest Grove OR 97116

EARLY

"Logging Tales Too Human to be Fiction"

With grateful thanks to so many who helped with memories, elucidations, corrections -- and spellings -- this book is dedicated to the courage of loggers and pioneers like Alfred and Hildur, Alfhild, Fred, and especially my Early.

EARLY

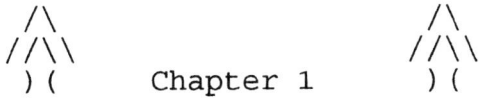

Chapter 1

When the tree ticks...

The sucker was thirty feet up. Earl, twenty-five in the early 1940's, Swedish blond and lumberjack muscled, dug in his calk (pronounced "cork") shoe spikes just under where the two by two-and-a-half foot sucker limb took off from the main trunk.

For a tree climber, the trick to keep from falling is for him to lean his weight into the rope encircling the tree trunk. Leaning back into it pulls its loop tight against the tree trunk opposite him. For safety, the rope is strung through the metal rings on his leather logging belt buckled around his waist. He balances on the cleats on his boots to keep from slipping down the bark, but the main force holding him is the tautness of the rope held tight against the bark on the other side of the trunk from the perch of his body.

He learns to hold that balance as he does the physical labor of trimming branches, sawing through the trunk, or pulling up equipment from his dangling second rope, the pass rope. It isn't like simply climbing the tree, where he clings and holds and pulls himself up. At ten, twenty, or fifty feet above the ground, he has to lean out away from the tree trunk so both hands are free to work. His climbing rope is literally his lifeline.

To move up the tree, he balances on his boot spikes, lifts the loop with outstretched arms and flings it up the trunk, then scrambles a few steps on his "cork shoes." In those moments he defies gravity with momentum. It takes practice to perfect the timing. Going straight up, the climb is comparatively easy.

But when Earl reached that broad sucker limb at thirty feet, he had a problem. He had to get his climbing rope to encircle the trunk above the sucker. The climbing rope is only about twelve feet long and wouldn't fit around the main trunk and the diverging sucker limb at the same time. Getting above the sucker is a trick of balance and movement that hopefully cheats gravity for longer than an instant.

Sliding his climbing rope through the ring on his belt to lengthen it to encircle the base of the sucker limb as well, Earl needed to get as high as the sucker would let him. He flipped and flung with the practiced motions of a veteran tree climber. He'd topped his first tree while still in his teens.

Now came the tricky part. He had to get back to encircling the main trunk. He reached up over the sucker limb, grabbing the far arc of his loop of rope, and pulling it toward him. That kept it tight enough against the two parts of the tree to hold him from falling back, while the "spurs" on his calk shoes kept him from falling down.

While holding that part of the circle with his left hand, Earl reached down with his right hand to untie the free end of the rope from his belt. The other, eye-tipped end of the rope remained secured to the belt rings, but loosening the free end meant that the life-saving circle of rope around the tree was broken except where he held it. He hung on. Thirty feet is a long way to fall.

With his right hand he pulled at the rope end that had been around the extra limb and was now hanging free behind it. He drew it up and over the crotch between the sucker and the main trunk and slid the free end back though the eye tip of the other end. He retied the loop into a complete circle again with the special sliding knot that lets you lengthen or shorten the loop, but which clamps down tighter as you lean your weight against it. His lifeline was intact again.

As a kid, he used to cling there for a moment, waiting for his heart to slow its anxious pounding, thirty or forty or sixty feet up -- wherever the sucker went out.

But now, after years of climbing around sucker limbs even eighty feet in the air, Earl merely tightened the rope loop to fit the tree above the sucker. Balancing on his spurs, he flung the rope up again and leaned back into it while he moved his feet to dig in higher up the trunk.

There is a rhythm to it. A good climber can fling the rope and scramble up, almost running up a tree. In a logging show, he can run up a hundred and fifty foot pole to ring a little bell at the top, then run down again, reversing his motions -- all in thirty seconds.

But this wasn't a logging show. This was Earl and his father Alfred Nordgren working their woods, the two of them miles from anyone else. They needed a spar tree on which to hang their overhead blocks in order to use lines strung through their pulleys to lift and move the felled trees.

Nowadays, logging outfits have a metal pole on a vehicle that they can drive to the chosen location, then raise the metal spar pole on its hinge and telescope it to stand high. But in the early days of logging, Earl and Alfred had to select a tall tree near the middle of their work area and prepare it.

The tree the Nordgrens had selected for their spar tree was in the center of the section they wanted to log, but it was a hundred and fifty feet tall, and they only needed it to be a hundred twenty feet. The branches or the whole extra thirty feet, if it were left up there, could snap off and tumble down into the multiple cable lines -- and possibly the men.

It was up to Earl to climb and "top" it, cutting off its top segment, before he could fit on the cable straps for the blocks and gear. That meant he had to chop off the highest thirty feet of this tree -- while his feet were a hundred and twenty feet above the ground and dug into the very tree he was chopping.

When the top of the tree splits and falls, it makes the remaining trunk sway in reaction, sometimes whipping so hard and so fast, the topper blacks out. All toppers quickly learn to secure themselves carefully before those final blows that may leave them dangling, unconscious and helpless.

In the nineteenth century, if a topper missed with his axe swaying up that high, he might chop through his own rope, cutting his loop. With nothing left to hold him against the tree, he'd fall. By the time Earl was topping trees, climbing ropes were made with slender steel cables incorporated within them so they wouldn't be so easy to chop through.

Earl climbed above the sucker. There were smaller and smaller limbs to clear, but Earl carried his axe and saw on his leather belt and hacked them off more and more easily on his way up. A spar tree must be bare of anything that might snag and foul the lines draping from its top.

Each time a severed limb was about to fall, Earl called down to warn his father on the ground. As each branch rumbled down from greater and greater heights, Alfred looked up, saw where the branch was coming down, and limped on his artificial leg to step aside, perpendicular to its path, out of its way.

This spar tree was five feet around, and it held its size for nearly all of the height that the Nordgrens needed. Delimbed and topped, the spar tree would then be strapped with cable bands to secure the four or five guy lines and then the pulley blocks through which to run the lines to lift and move the felled timber below.

The topper would also rig a straw line for lifting a logger up to work on the rigging, without his having to climb with "cork shoes" and climbing rope. He would fit circles of chain around his legs and be lifted up by a steam or gas boiler, called the "donkey engine", on the straw line that dangled from the its pulley near the top of the spar tree. It was safer and faster to be hoisted up rather than having to climb up each time a block needed to be moved or work done on the lines.

Earl had a lot of heavy physical work left to do at 120 feet in the air, but his first order of business was to top off the highest extra thirty feet of trunk of what, even this high, was still a thick-trunked tree.

As Earl started chopping into the trunk, he cocked his head. His axe blows sounded strange.

The tree had a shell of only about two inches of good wood to hold it together. The rest inside was rotten.

While the shell would still be strong enough to work as a spar tree, Earl knew from experience that in topping it, the shell could split down its length, or "slab out." As the tree split, it could sever Earl's loop of climbing rope and send him plunging 120 foot.

"It's a shell, Dad," Earl called down. "Let's blast off the top."

One good reason for using dynamite to blast off the top was that the tree was less likely to slab out. But an even more enticing reason to a topper, was the fact that he could light a long fuse and scramble down to the ground, out of harm's way before the top was blown off.

Alfred signalled that he understood. He had supervised the digging of railroad tunnels before he came to the Oregon woods, so he knew the ins and outs of dynamiting. He wired some sticks of the explosive together and, with a fuse and cap, attached them to the pass rope for Earl to haul up.

Earl strung the dynamite necklace around the tree at one hundred and twenty feet. Securing it, he let himself down the trunk about twenty feet to light the end of the fuse, then scrambled on down to the sucker. He climbed laboriously around it, reattached his rope around the tree trunk below the sucker and climbed on down to the ground.

While they waited for the fuse to burn its way up to ignite the dynamite, Earl and Alfred shared their lunch sitting on the ground some distance from that tree. It seemed to take quite a long time for the fuse to burn, but it finally did. Baroom!

But the blast only shook the tree. When they looked up, the unwanted top thirty feet of trunk and branches were still up there. The top hadn't been blown off after all.

They finished their sandwiches and thermos of coffee, and Earl leaned back to stare up at the spar tree.

"I better go back up and set another, larger charge this time," he said finally, and Alfred nodded.

Reloading his gear belt, Earl climbed back up. It was easier this time with most of the extra branches already cleared, but that huge sucker limb remained. At thirty feet, he worked the rope, swinging up around the turn and the base of that sucker, then reaching through and holding the back arc of rope while he untied the end of the loop at his belt and yanked the rope over the sucker and retied his loop around the upper trunk again. He climbed on up, much more quickly this time.

He got to the 120 foot mark and clamped in hard. He was leaning into his rope so he could get the new string of dynamite secured when he thought he heard something. Stopping the quick movement of his hands, he ducked closer to listen.

"Tick."

Earl frowned. Then he heard it again.

"Tick, tick." Then faster, "tick, tick, tick,tick, tickticktick..."

He sucked in his breath, realizing the first dynamite charge had done its job after all. He hollered down, "Dad, I think this thing's gonna go!"

He hurtled down that tree at a pace that would have drawn notice even at a logger's exhibition -- until he got to that big sucker. He couldn't get around it fast enough. And it was thirty feet up -- too far to jump. He clung there and the tree, up where it had been blasted, broke off. The top started down.

"Look out, Dad! It's coming right at you!" Earl screamed, unable to help his father or himself.

Alfred started stumble-running on his artificial leg on the uneven ground. But instead of running perpendicular, to one side away from where the tree was falling, he ran right out along the same way that the huge top was coming down. It crashed down past Earl and right at Alfred.

Working his rope frantically, Earl maneuvered around that sucker and got down to the ground. He scrambled out beside the top trunk, still three-and-a-half feet around, though it had come from over a hundred feet above them.

"Dad! Dad, are you all right?"

Crushed and twisted branches surrounded Alfred, within a foot or two in front and behind. He was scratched and bleeding, and shaking. But the top trunk had missed him completely. Alfred needed some help to stand, but he shook his head to tell his son he was all right.

When he could talk again, he swallowed, gulping, "Oh, Early, don't let's tell anybody about this -- especially your mother."

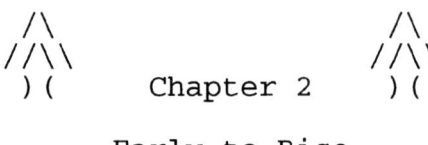

Chapter 2

Early to Rise

"Early! Early, get up -- up and jump on the stump!"

Little Earl, youngest of the three Nordgren children, poked his nose from under the quilt on his cot in the attic.

"Yeah, I'm coming, I'm up, Ma!" he called, then snuggled back into the warmth of his nest, knowing how cold the unheated attic floor would be when he did finally have to drag himself out of bed. Their mother had wrapped a hot flat iron in newspapers to slide under the covers to warm their beds for them before they got in last night, but there was nothing to chase the chill yet this morning. Though the chimney reared up through the attic to the pitched roof, it gave barely enough heat to keep dry the enormous, once-a-year-purchase sack of sugar propped beside it.

The length of the attic was the entire width of the wooden house their father, Alfred Nordgren, had built by hand on their homestead high in the glen near the top of Carpenter Creek. Earl Marvin slept at one end, next to his brother Fred (really Manfred, but Fred hated that name, and he was big enough that nobody called him that more than once.)

At the other end of the attic, their older sister Alfhild had her own cot and night table and the one kerosene lamp. Her father had named his first-born by combining his name with that of his Swedish bride, Hildur.

Fred turned over without much of an answer to his mother's call. Even Alfhild, sweet-natured and usually conscientious, was slow to get up. Without central heating, the house was warm only by the woodstove downstairs in the kitchen or the wood heater in the livingroom, far away from the children's snuggly beds.

Alfred usually left a small fire going in the heater, but that often burned out during the night. On cold mornings, it was a chore for the first person up to hurry and get the fire restarted. Then most of the family huddled around it for a while before they could get their blood circulating again.

But the chores had to be done every morning, and Father made sure they were.

Fred and Earl gathered the hens' eggs and milked the cows before breakfast, which the family always sat and ate together. Alfhild helped her mother prepare the huge pan of hot oatmeal mush on the wrought iron woodstove in the kitchen. Earl got so tired of oatmeal mush, he never ate it again once he'd left home.

> As Hildur cooked, she sang,
>
>> "Many fishes in the brook,
>> Poppa catch 'em with a hook,
>> Momma fry 'em in a pan,
>> Early eat 'em like a man."

The woodstove was large, with an oven to one side below and a warming oven above the flat surface with its heavy burners, which could be lifted out to stuff wood down into the fire. Pipes coursed through the back of the stove, heating their water. But with the fire out much of the night, water for washing in the morning seldom had lost its chill.

As the heavy breakfast concluded and Hildur and Alfhild began clearing the table, Alfred poured some of the coffee from his cup into the saucer. He slipped a sugar cube into his mouth and sipped from his saucer, sieving the hot, black liquid through the sugar cube in the ancient Scandinavian custom.

Breakfast over and chores completed, the three children walked the mile and a half to the one-room school house Alfred had helped to instigate and build. Maxine Hoover Schaefer's father, Nat Hoover, had donated the land, and Alfred gave much of the work and wood to build the schoolhouse.

Sometimes Father had to break trail through the snow for them with one of their two old white workhorses, Pat and Mike. It was tough to get to school some days, but Father always saw to it that the three Nordgrens got there.

Alfred Nordgren had emigrated from Sweden as a very young man to find work. He was a tunnel builder for the railroad, and he was good at it. On one job he had the first crew start on one side of a mountain, and the other crew dig in from the other side. When they broke through, their tunnels were only off a foot and a half from lining up perfectly.

Alfred had had a promise of marriage from Hildur's older sister back in Sweden. But the long wait for him gave the young woman time to dwell on the horrors of moving so far from home and family. Finally she told Alfred that she wished to marry another young man -- one who would stay in Sweden.

Her younger sister Hildur had been farmed out to live with an aunt when she was just a tyke. Times were tough in Sweden, and there were just too many in the Thorin (Anglicized to Turin) family for Hildur to be able to remain with her twin brother and other siblings.

When Alfred sailed back to Sweden, his betrothed was marrying the other man, so he asked Hildur instead. She agreed, and accompanied him back across the ocean. Her wide gold wedding band is inscribed: "A. H. 8-22-08"

Hildur knew very little English when she came to North America. At first she worked for a family, cleaning and cooking. They asked her one day to make them a pumpkin pie, which is not a common treat in Sweden. In fact, Hildur had never before prepared a pumpkin in any way. She did what came natural for her specialty, apple pie. She pared off the tough outer rind and scooped out the seeds, then sliced the pumpkin into the pie shell like apple slices. It was an interesting pie, but no one asked for seconds.

Alfred and Hildur's daughter Alfhild was born in Canada. The family moved then to Spokane, Washington. Alfred was growing sick of the corruption and crooked business practices in tunnel building. He wanted to give it up and get a place of his own where his family would grow up around horses and cows and chickens and the wholesomeness of basic values. So he bought the only land he could afford -- cheap.

Alfred and Hildur became homesteaders up on Carpenter Creek in western Washington County, west of the Cascade Mountains, but on the east side of the Coastal Range in northwestern Oregon.

Manfred was born near Carpenter Creek about three years after Alfhild. By the time Earl came along another three years later on Valentine's Day of 1917, Alfred had built the home place. It was up the road about a quarter of a mile from the log home Alfred had built for his friends, the Hills. The Nordgren homestead was higher up in the glen carved by tiny Carpenter Creek. As the Nordgrens moved into their home place, Alfred built the barn for the work horses and cows and Hildur's chickens.

It is doubtful that Hildur had any say about their move from Canada or from

Spokane or their purchase of the homestead property in the wilds west of Portland. It was the Old Country way that the man of the family made the decisions without discussion or consultation. The woman was expected to accept and be grateful. (Many years later, Alfred bought a large, two-story, frame house on 17th Avenue in Forest Grove and shopped alone for furniture, then brought Hildur to the furnished new home she'd never picked or had a say in. She loved it.)

The new Nordgren homestead deep in the woods up Carpenter Creek had a huge pantry off the kitchen. Hildur bought food in bulk and stored it in that pantry, like the one foot by two foot box of raisins that the puppy got into. That puppy ate raisins and ate raisins, until he blew up like a balloon. He could hardly waddle out of the pantry, and it was touch and go for a while, but he lived.

Other attractive treasures were also stored in that pantry, like the German Luger pistol Father kept on a high shelf. Fred climbed up and stole that pistol down a few times. He even took it outside and shot it, but even Fred only did that once or twice.

Hildur stored Father's Copenhagen chewing tobacco in the pantry, too. Fred stole some one time and took it to the one-room school house. He and neighbor boy Donald Brown tried to chew some of it but turned green and sick as dogs. They lived, but for a while they weren't sure they wanted to.

The home place had an attached woodshed, which had to be kept stacked with wood all the time because wood was their only means of heating and cooking.

One time Alfhild was left in charge of seeing to it that her brothers swept the woodshed clean. Well, the boys did sweep, but they piled the sweepings behind the chopping block where Alfhild couldn't see the dirt. She suspected they'd been up to something, though, and as she made their noon meal, she got even. She put castor oil in their lunch.

There is still a saying in the family: "Faster than castor oil through a tall Swede."

The children had only one lamp in the attic and Alfhild usually had it by her bed so she could read. Sometimes when she fell asleep, Fred or Earl would sneak over to her end of the attic and carry it to their cots. But Alfhild always woke up and came over to take it back. One night Alfhild blew out the lamp and turned over. It was only 9:30, but the whole hard-working country family was asleep by that time.

This once, though, the wick hadn't quite extinguished. It smoldered for some time with the kerosene level so low that the end of the wick wasn't really in it any more. When the charred tip of smoldering wick dropped off from above, it fell into the kerosene, setting it ablaze. The lamp blew up like a torch, whooshing and flaming so the children, startling awake, thought it must be a Roman candle. They struggled to sit up in their cots, too frightened to move until Earl leaped out of his bed and ran to Alfhild. He scooped up the lamp and hurled it out the attic window, clear over the woodshed and down the hill. It hit the path below, shattering and exploding like a Molotov cocktail.

Alfred came roaring up the stairs to find out what was going on. When Fred and Alfhild told him what Earl had done, their father went pale. He knew such a blaze might well have burned down the whole house they'd worked so hard to build.

It wasn't all work on the homestead. Father sometimes dammed up trickling

Carpenter Creek to make a swimming hole for his children.

Earl tells a story of how he learned to swim: "They threw me into the middle of the lake, and I had to learn to doggie paddle or I'd'a drowned. That wasn't so bad, getting to shore, but it was tough getting out of that sack they'd tied me in."

Those who might suspect some exaggeration in Earl's telling point out that the swimming hole, even in the wettest summers, was never big enough for more than one kid to wet his kneecaps at one time. But it was fun. Especially when their father let them get Pat, the huge, long-legged old, white workhorse from the barn and back him up next to the edge of their pond. Alfred's other workhorse, Mike, had died, but gentle old Pat would stand patiently on the bank and let the children scramble onto his back and slide off his rump into the water.

Earl was particularly fond of Pat. Being the youngest of both his family and of the neighboring Brown family of four boys, Earl was used to being picked on. But he didn't like it. When the teasing got to be too much, Earl would dive under Pat's sagging belly and sit there where the other kids wouldn't go under after him.

The Nordgrens only got down the hills into the small town of Forest Grove once a week, on Saturdays. On the Fourth of July, they went all the way to Hillsboro -- ten miles! The celebrations were held where the Hillsboro library and Shute Park are now. A carnival was there some years, and fireworks, and a dance afterward.

One of the carnival attractions was a guy on a motorcycle who went round in a huge cage and out onto a long ramp. He'd gun his motorcycle and roar up the ramp and jump over cars. One year he didn't make it all the way over the line of cars.

But even years when there wasn't a tragedy to mar the celebration, the pull of work on the farm never let the Nordgrens stray far from home for long.

"Sometimes Dad wouldn't let us stay for the fireworks," Earl says. "We had to go home and milk those cows. Boy, that went over big. I still hate cows."

If getting the ten miles to Hillsboro was a momentous occasion, going to Portland -- nearly thirty miles due east of Carpenter Creek -- was a once-in-a-childhood event.

"When I was kid we went to Portland once -- and that was it. We stayed at a hotel overnight. Mom and Dad went to the movies. They took Alfhild with them, but Fred and I had to go to bed.

"'How come Alfhild can go and we can't?' we whined.

"'Because she's older,' Mother tried to explain. 'When you're older, you can come, too.'

"Fred looked at me and I looked at him, and we both were so mad we wanted to cry or fight, but we knew we didn't dare.

"Finally Fred exploded, 'It's not fair! We'll never be as old as Alfhild!'"

The day the elegant, high Longview-Kelso bridge was opened across the Columbia River between Oregon and Washington, Earl, Fred, Alfhild and their mother went to see the historic event with a friend in his shiny Nash car. Even Alfhild was a little girl then. Earl doesn't remember the trip at all, but Alfhild remembers the thrill of the huge car. That thrill wore a thin as they waited on Oregon Route 30 for hours and hours before the opening ceremonies.

Hildur had made them a picnic, and that helped make the long wait much more festive. At least until the ham and the pickles and deviled eggs and homemade jams and homemade bread and pies and cookies were spread out on the blanket. It was all ruined. Hildur had forgotten the butter.

In the old days of radio, the Richland News ended up with a deep-voiced man saying, "What a world!" But that was when the Nordgrens finally got electricity. For most of Earl's growing up, everything was done by hand, including washing the clothes for the growing, active family.

Hildur had been scrubbing overalls and shirts and skirts and dresses on a washboard for many years. When Earl was about eleven, he and Fred and Alfhild and Alfhild's young man, Charlie Zumwalt, all chipped in together and bought a gas-powered washing machine. They had it delivered and set up out there in the country. When Hildur saw what it was, she was overcome with emotion, something the children had very seldom seen in her.

"It's a life-saver," she cried, in tears.

One Christmas, Earl bought his parents a Seth Thomas clock for the mantelpiece, a sleek, wooden clock with low wings sweeping up each side to surround the round, ivory-white clock face. Proudly, he showed them how to wind up its works. Many nights he lay in the attic listening to it chime the hours of darkness.

"At the farm, we didn't have air conditioning, of course, and the flies were so thick. There were so many flies, they were always all over you wherever you went or whatever you did, especially if you'd worked up a sweat." There were no screens on the windows, and the dangling sticky fly paper only rid the house of so many of the buzzing, black insects. "Dad always used to say, 'You can't fool the flies.'"

Fred told the story about the boy who bragged, "My dad can lift a hundred pounds with one arm." The boast was answered: "Oh, you think that's something? My pop has gas on his stomach."

Alfred gargled with Listerine every day. One day he got the wrong bottle and gargled lineament. "He did have a time!"

"When I was little," Earl remembers, "when somebody tickled the soles of your feet, they'd say, 'Tickle-lickle-lickle-lickle.' When we hit our elbow, we'd say, 'I hit my crazy bone. Now I'll have to go and comb my hair again.'

"Are most of our memories tied to emotion?" Earl remembers as a three-year-old that he'd had a bubble pipe. He lost it. "I looked everywhere for it. Eighty years later, I'm still looking for it."

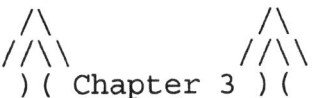

Chapter 3

Look To The Hills

"When the folks first started homesteading on Carpenter Creek," Earl mused, "the trees were so thick Dad said you could walk in at one end of the canyon and never see daylight until you came out the other end."

Alfred cleared some of those trees to build a log house for his friends, the Hills, then his own home place and barn higher up in the canyon. Years later, to support his family in the Great Depression, he hiked up into his forest to cut firewood to haul down into Forest Grove to sell by the cord.

Gradually Alfred realized that the economic times weren't getting better any time soon and that he'd do better financially if he cut timber, not just firewood. So he learned to be a logger. Fred and Earl learned with him as boys growing up.

Most of the work was done by calloused hand, with power generated by straining muscles in arms and legs and back. When the Nordgrens found they needed road tracts through the woods on which to haul out the timber, they hacked out over five miles of road up into the Coastal Range, sweating over pick and axe and shovel.

Alfred had used a lot of dynamite in blasting out tunnels for the railroad. He used it now to blast out tree stumps they couldn't dig under or haul out to clear for their roads. For one enormous stump --

four foot across -- Alfred had to use a whole box of dynamite. Finally it blew out. There were scars on the surrounding trees where pieces of stump sailed through, and no one could find anything left of that stump. But they got their road through.

When Earl was nearly nine, Alfred had bought a new car, but he seldom took it up onto their hand carved roads.

"I remember the smell of my dad's new car when he bought it about 1926. Nothing like it. That car was about the most marvelous thing I'd ever seen. It even had tail lights!"

As children, Fred and Earl leaned out the side of the car as Alfred drove all of 10 mph. The boys saw a man sitting at the edge of the road with a pipe in this hand. That pipe was just too much of a temptation for the two, who leaned far out with a cane and swiped the pipe right out of his hand as they drove by. Seventy years later,

Earl remembers the startled expression on the man's face.

In the woods, Alfred used his 1919 hard-tired Master truck, that Alfred once got going up to 35 mph, and somebody passed them! It wasn't comfortable going that fast, though, as there was not much in the way of springs, and the rutted road made the trip so rough it felt as though your teeth would be shaken out.

Alfred hired neighbor Carl Brown to drive the truck. The Browns lived on the hill where the Vaanderings lived, and their sons -- Lell (Lewellyn), Ted, Donald and Carlton Brown -- went to the same one room schoolhouse as Alfhild and Fred and Earl.

"I've never tried that painting or sculpturing, high class stuff," Earl laughs. "Where I grew up, elementary school was considered higher education."

There was no school at all in the area when the Nordgrens moved in, so Alfred and some of the neighbors got together to build one on land that Nat Hoover donated.

When Earl graduated from eighth grade, he was the head of his class. He was also the only one in his class so he guesses that made him the tail of the class, too. During most of his childhood, there were as many as eight kids in the whole school at any one time: Alfhild, Fred, and Earl Nordgren, the four Brown boys, Maxine Hoover, and the three Person kids.

Maxine's dad Nat Hoover worked at one time for Alfred Nordgren, whom he called "the old gent."

There are stories that Maxine used to throw tantrums as a child. She'd fling her doll under the bed, and her mother would sigh and get down on her hands and knees and crawl under the bed to get it. Maxine, smiling impishly, is supposed to have commented, "I sure get a kick out of that old lady."

After Maxine grew up to marry Lyle Schaefer, there were no more tricks with dolls.

The oldest of the Brown boys, Lell, met a lovely lady named Alma in Eugene, Oregon. As they talked, he realized this was the little Person girl whose family had moved away from Carpenter Creek some fifty years earlier when she was just a child. Lell and Alma were married not long after that long-delayed reunion.

Ted Brown -- whose pretend-vehement "Ju-das priest!" tells you where he is even when you can't see him -- has been a dear friend for eighty years. As Earl tells it, Ted was always pretty quick at figuring things out.

"One time Alfhild was driving down the dirt road in our old Chevrolet touring car with side curtains. Ted was driving up the hill when he saw her. Cars only had two wheel brakes in those days and they were always out of line. So Alfhild really only had one wheel brake, and she was standing on it. But the Chevy was gaining speed faster and faster as she went down the hill.

"Ted saw her coming and realized what was happening. He jammed his car into reverse and backed down that hill fast enough in front of her that she never did hit him."

Ted wasn't the only one in Carpenter Creek who could move quickly when he had to. "We knew what a still was," one old-timer says. "We had two up Carpenter Creek, one on land near the Browns, and another one higher up in the canyon. The Revenuers came sometimes, but there was always a tip-off ahead of time so those stills got all tore down and hid before they came."

There were a few other neighbors up Carpenter Creek. Swanson was an older man with asthma whose wife was gone. His breathing problems got worse and worse until finally he had to see the doctor. He came home with prescription pills to be taken three times a day for a week. But, perhaps figuring if one was good, more would be better, the old man took them all -- all at once. He slept for three days.

Alfred missed seeing Swanson that first day, but there was plenty to keep the Nordgrens busy with their own work. When no one had seen him the second day either, though, Alfred got concerned. By the third day he was worried enough to trudge up the hill.

Old Man Swanson's cows were wandering around, mooing piteously. They hadn't been milked for three days and they were in pain. In the house, Swanson was snoring loudly. Leaving him in bed, Alfred went out to round up the cows and milk them. He got the farm back to rights and waited for the man to finally wake up. The story goes that that unorthodox use of the prescription did something so Swanson was never again bothered by asthma. Or perhaps he knew better than to complain about it, because there was no way he would take those pills again, even the way they'd been prescribed.

It had been raining long and hard, even for Oregon, and another farmer neighbor farther up Carpenter Creek had laid out planks, end to end, to cross his yard, which was ankle-deep mud. He was startled awake one night by noises in his chicken house. Hearing the hens cackling and flapping around, the farmer rose up and hurried across the yard on the boards. He was so intent on keeping his balance running on those planks that it wasn't until he got to the coop that he realized what was causing all the commotion.

The skunk took exception to his interruption and sprayed the farmer, direct hit.

Yelling, the hapless man stumbled back across the wood planks and up to the house, but his wife was awake by this time. She smelled him coming and locked the door. No way was she going to let him into the house. He had to sleep in the barn.

Another canyon neighbor's wife sent her man down to borrow money from Alfred and Hildur to make payment on his life insurance. She was afraid of what would happen to her if something happened to him, but they didn't have any money to keep up the policy. Alfred didn't have any money either, but he gave them a calf to sell. The neighbor took the calf and sold it and kept up the policy -- then his wife died before he did.

Dances up in Carpenter Creek on a Saturday night could be pretty lively.

Neighbors came from way back up in the hills. Whispering Smith used to play "Pretty Redwing" on his fiddle. It was the only tune he knew so he'd keep playing it over and over. But the folks were drinking and laughing, and nobody minded because no one else could play anything at all. Whispering Smith played all night long and was invited back for every dance.

Allard played an accordion. He wasn't from Carpenter Creek; he worked for a Ford assembly plant in Portland. But when the people found out he played an accordion, he got invited to all their dances. Allard probably weighed two hundred and fifty pounds. So did his wife. One night they came up the hill in their old Ford car, bringing Allard's sister, who was almost as heavy as he was. And they brought their son. He was only 200 pounds. He was young yet.

Allard played his accordion and everybody laughed and danced and drank. Later, after everybody was starting home, Allard and his family went out and climbed into their car. Those wheels just sighed and spread apart. They kept spreading wider and wider until Allard and his wife and son and sister were all sitting on the ground. They just laughed. They stayed overnight with some people at the dance. Next day, Allard went into town and bought a new Model-T Ford.

    There were branches into Carpenter Creek above the Nordgren home place that had carved out their own glens. Coosman Oginovitch lived up one of those side glens. He used to say some of those canyons were pretty good. You could holler up, "What time is it?" and the echo would come back after a while with, "Two o'clock."

Coosman was a homesteader, too. He hiked back up into the hills and felled trees and cleared and put up his own cabin high on the creek branch. It was fifteen years later that he found out he'd built on the wrong spot. That wasn't even his land.

Coosman built his cabin between steep walls in a very narrow canyon so the cabin was almost right into both walls. The creek ran right underneath. On the first level, he kept goats. On the second level up was the kitchen, and on the highest, his bedroom.

One tree Coosman was felling above his place crashed down and got away from him. It landed right through the roof and came to rest on the floor of his bedroom. So he carried his tools into the cabin up to the third floor and sat on the edge of his bed calmly sawing the big log into cord wood.

Coosman was excited as he told Jake Tohalsky about the most beautiful black kitty he'd ever seen. As Jake listened to the description, he finally asked Coosman if that pretty kitty had a bushy tail and a white stripe down its back. Coosman owned as how it did.

"Now that there isn't a kitty, it's a skunk, and you better stay away from it," Jake said.

Coosman shook his head. He wasn't about to be afraid of such a pretty animal, and besides, it had gotten quite tame. It would come almost right up to him sometimes.

"Well, you best keep clear of that thing, but if you won't listen to me, just you yank that thing by the tail if you want to find out why I'm warning you," Jake said, exasperated.

Next time Coosman came across that pretty kitty in the woods above his place, he reached out and yanked on that tail. As he told Jake later, "That kitty stinken up whole valley. Me take him up high in kinyanyon and gravel him over."

Jake Tohalsky had his own cabin on Carpenter Creek. One day his wife was working in their kitchen while he was up in the woods above their cabin, cutting 16 inch sections from logs he'd felled. One of the sections got away from him and rolled like a wheel down the canyon.

Pitching his axe, Jake took off running after it, but it was picking up speed with every turn, rolling down, faster and faster, heading straight for the Tohalsky cabin. Jake yelled to warn his wife. He kept hoping the runaway section would bound against a rock or a tree and be diverted from the house, but nothing seemed to be in its way. It was moving like a ball bearing heading for a powerful magnet.

Jake screamed as the log section hurtled into the back wall of his house. It made a hole and disappeared inside.

When Jake finally stumbled in at the door, Mrs. Tohalsky stood trembling in the kitchen, staring at him. Her eyes followed the path that log piece had taken coming at the back wall and straight through right behind her as she stood at the sink, then out through a new hole in their front wall.

Jake didn't say much. What was there to say?

Neither did Jake find much to say the day he was carrying a heavy pulley slung over his shoulder up the hill in the Nordgrens' woods. When he got to where Alfred and the other loggers had gathered, he dipped his shoulder and whipped the pulley forward to set it on the ground. But he had his finger jammed into a niche in the pulley. The metal didn't give, but Jake's finger did. As Alfred helped him extract it, Jake stood up slowly and looked at white bone ends stuck out from torn skin and sighed. "I think I broke my finger," was all Jake said.

By the turn of the twentieth century, Carpenter Creek residents still had fairly large areas of unspoiled country that proved irresistible to a number of city folks who drive into the hills, ignoring 'No Trespassing' signs and closed gates. Locks are often found broken.

A large group of men in blue jeans and baseball caps were drinking around a campfire high in the Carpenter Creek woods one dry summer evening. They looked up as two women approached cautiously.

"That's quite a camp fire," the older woman commented. "We were told it's too dangerous for any fires this late in a dry season."

"No, it's okay," one of the men assured her. "We got permission from the Forestry Department." Several of the others grinned and looked down.

"You know, this is private property," the woman tried again to reason with them.

"Yeah, yeah, but we got permission to be here from the owners."

"I don't think so," she shook her head. "I'm the owner."

Fire is such a hazard in the heavily wooded hills, all landowners find themselves anxiously holding their breath from late summer until the rains come, especially in a dry year.

Lori and Ivan Andresen bought land from Earl in the early 1970's, with Bill Franck as their agent and ultimately their friend. They tramped up into the tree-covered hill to look, though Ivan was recovering from surgery. He knew he had to see for himself the lay of the land. Lori kept watching him furtively, hoping the climb wouldn't be too much for him.

But it was Bill Franck who called a halt, breathing hard.

"Give me a minute," he panted. "I just had heart bypass surgery, and I'll have to rest a bit here before we finish the climb."

"Are you sure you can?" Lori's eyes widened. She knew she couldn't carry them both. "Let's just go back down to the car."

"Oh, no," he chuckled. "I'll be fine."

When they'd panted and grunted their way to the knoll, they saw spread out before them the gorgeous line of the hills just west of Portland, with magnificent, snow-capped Mount Hood beyond them to the east. The Andresens knew right then that this rugged acreage would be home.

They strung tape for their winding driveway to climb the hill through the trees, measuring carefully 30 degree curves that would accommodate hauling their 64 foot trailer to the vista knoll.

They were building their house, set to take in the panorama of mountains. They'd put in uncounted hours that hot, dry summer, toiling to put in the long driveway and the foundation and superstructure for their lovely home with its daylight basement. At first Ivan, sweaty and tired, couldn't register what his son Matt was telling him.

"Daddy! Smoke!"

It wasn't to the dense stand of trees south and west of the house that his little boy was pointing, but to the north, toward their David Hill Road neighbors' rolling hills covered by waist-high grass. North?

But there wasn't anything to the north that should catch fire, nothing that could possibly send up that kind of billowing black, dense smoke. A fire like that would be sure to light up the tall, dry grass. And the grass would fuel the fire all the way up to the Andresens'. Ivan knew in the instant he looked at that terrible smoke that they could well lose everything they had worked so hard to build.

The volunteer firemen were there within minutes. The long, winding, unpaved drive with its gentle curves accommodated their firetrucks as well as it had the Andresens' long trailer. No one could believe the black anger of that fire raging in the glen just north of Andresens' property -- until they finally got it under control. In the bottom of the glen were the charred remains of a Pontiac car.

There are trees beside David Hill Road now, but at that time the car owner simply drove off the road, across the field of tall grasses to the rim of the glen. He stopped there and got out to pour gasoline onto the car and wedge the accelerator. Then he lit the car on fire and sent it over the edge into the canyon to burn.

He'd reported it stolen. He wanted the insurance money.

The fact that in August he could have set the entire hillside ablaze didn't matter to him as much as a chance to get that insurance money.

But the Andresens' next close call with fire in a dry August didn't make even that much sense. Young Matt, who was just beginning to drive on his own, was on his way down their hill to town when he swung through to the last of those gentle curves -- and came upon a tinny camping trailer parked at the first clear landing halfway up their driveway. A rugged man, woman and children were gathered at their portable table. Their campfire blazed unrestricted beside the Andresens' woods.

"Dad," Matt phoned from a neighbor's home at the bottom of their hill, "I tried to tell them it was private property and that it was much too dangerous for them to have a fire, but he gave me a pretty rough time. I told him he'd better clear out because he didn't want to deal with my dad."

Ivan scooped up his 12 gauge shotgun and headed down his driveway. The intruders had taken off in a hurry. They'd kicked the fire "out," but cedar logs, scattered into the dirt of the drive, smoldered and shot sparks into the Andresens' timber. The trespassers must have brought the logs with them to fuel their campfire. The Andresens don't have cedar on their property.

Lori, Ivan and Matt worked a long time stamping and watering down those cedar logs until they could be sure all fire danger was past.

As Earl gradually accumulated 5000 acres of timberland in the foothills of the Coast Range, summers were a constant emotional drain as he worried about fire.

Most summer evenings he sat in his yard looking west for rising smoke or the dreaded red glow that would signal a fire that could consume everything he owned.

In the catastrophic Tillamook Burn, fire had twice come to the edge of Alfred Nordgren's timber property and stopped for no obvious reason. Earl was as grateful as his father had been, but that kind of extraordinary luck was not something any man could count on.

The property brought other problems as well. One summer, as Earl checked his roads, he kept finding dumped garbage. Some evenings he drove up into his woods and parked, just waiting to see if he could find the guy who was hauling his trash out and dumping it. One night a pickup did come, with a young man behind the wheel and a small boy on the seat beside him. The driver stopped and got out, and was hefting the bags of garbage from the back of his truck when Earl appeared from between the trees.

"Is this really what you want to teach that little son of yours?" Earl asked.

The young man shook his head slowly, dropped the bags into the bed of his truck and backed away. Earl had very little trouble with dumping after that.

Art Vaandering was driving home past the Nordgren's old log cabin one night when he heard noises. He parked and hiked up to have a look-see. A dozen or so kids were in the cabin, probably students at Pacific University in Forest Grove. They had a keg of beer in the middle of the floor from which each was tapping liberally.

Art hiked back down to his pickup and went to telephone Earl. "Got a bunch of college guys up in the cabin, Earl. Did you give them permission to party there? No? I didn't think so. No, don't worry,

I'll take care of them for you," Art smiled and hung up.

He dug among his tools for his special portable light and drove with it up into the woods near the cabin. Setting the light on top of the truck, he switched it on. Bright red beams swished and flared. Bodies flung themselves out of every conceivable exit from the cabin, except the front door. Young men crashed and scrambled through the dark woods. In moments, the cabin was empty, except for the abandoned beer keg.

Art waited a while, but when no one came back, he smiled at the keg. "Looks like they left you in the lurch, old fella. Come on, I'll take you home with me."

Monday morning, two of the college men came to Earl, wanting their beer keg back. They had a large deposit on it and stood to have to pay more than $200 if they couldn't return it.

"Well," Earl said, looking at them, "you fellows did quite a bit of damage, you know. Tell you what, give me $100 to fix the windows and the other damage, and I'll give you back that keg."

They hung their heads. There was no way they could come up with the money, though they offered to work it off.

"Well, now, that's okay," Earl said. "I'll just talk this over with the president of the college and see what he thinks about how to handle this. He's a good friend of mine, you know."

The two young men stared at each other and were gone. Within an hour they were back -- with $100 in one dollar bills and change.

Chapter 4

Keep your chin up -- it helps keep your mouth shut

Bill Lepschat needed to make some money one summer during the Great Depression. There weren't a lot of options, but he did find a local cannery that was contracting for farmers to grow beans for them.

Bill signed with them, bought the bean seed from them on credit and planted when and where they told him, on his land that he'd plowed the way they'd instructed. He was told how and when to hoe, what fertilizer to buy from them and how and when to use it. When the crop came in, he harvested when and how they said.

And when he got done with all that work on his own land, he only owed them $600.

Earl hated farming. The cows the littlest Nordgren boy had to try to bring home from back up in the hills might have had something to do with that. They were notorious for hiding up among the trees, standing stock still when they heard him coming so the bells around their necks wouldn't ring. It was unnerving enough to be alone high in the hills searching for those ornery creatures, but, more than once, little Earl heard coyotes nearer in the dark than the family bovines.

His father wasn't really much of a farmer either. For Alfred, the homestead was more of a means to stay self-sufficient. He'd emigrated from Sweden when there simply wasn't a way to earn his living in his own country. He wanted his children raised in the rural way of life, and he wanted always to be able to supply at least their basic needs. When they needed income, he and the boys carried axes and saws up into their forest to cut fire wood.

Alfred always gave full measure, plus a little more, like adding cross pieces so the ends of the stack would not roll. So he was dismayed when "V" called to complain that he'd been cheated. Shaking his head, Alfred took more wood and filled in the chinks as V demanded.

Later V's neighbor came out to tell Alfred that he'd seen V slip wood out every other one from the stack and carry it inside before he'd called to complain.

Alfred got a contract to supply a large amount of wood for the college in Forest Grove. It was a good arrangement for both sides. Alfred had the security of a large contract. Pacific was getting full cords of good wood, delivered on time, as agreed.

At the end of the season, Alfred went to talk with the man in charge of renewing the deal. Alfred was welcome to next year's contract -- IF he made sure a few extra cords of wood appeared for free on the man's back porch for his personal use.

Alfred had quit the tunnel-making business for the railroad because of the crooked dealings and politicking. He wasn't about to get suckered into that same kind of dealing at his newly established home. He wouldn't agree to the payoff.

"Suit yourself," the man shrugged. Alfred did not get the college contract the next year, or ever again.

The Nordgrens continued selling cords of firewood, but without the security of a large contract, it was piecemeal at best. The Great Depression was sapping everyone's strength.

Earl remembers helping take cords of wood to deliver in Portland. The Nordgrens made the trip via planked Canyon Road. The cords were supposed to sell for $4 each, but sometimes after they'd hauled it in and stacked it, they wouldn't get paid.

One family told them they intended to pay, but they didn't know where their checkbook was.

Their little girl's face brightened. "I know where it is!" she cried and ran inside to get it.

Alfred was grateful he'd secured the homestead for his family. They would always have enough to eat because they could raise it themselves. And he had the forest on his land. He looked again at the thick stands of trees. He was beginning to realize there was a great deal more money to be made in logging than in cutting fire wood.

Alfred had no experience in logging, but he was an able man, willing to learn. His two young sons, Fred and Earl, were learning with him, all but absorbing the hard lessons Alfred garnered from each mistake. Both sons became quite skilled in logging.

"I could throw my cap out at the top and beat it to the bottom if the tree wasn't too high -- maybe 60 feet or so," Earl said, but he knew there were many others who could run up and down the trees even faster than he could.

Logging for someone else was hard, dangerous work, and not designed to make the logger rich. In 1929, a hooker (or choke setter) was earning about 40 cents an hour. By 1946, he was making $1.25 an hour. The whistle punk made less than half that much, and the water boy got about half what the whistle punk made.

The Lyda brothers bought stumpage in World War II, at $1.50 per 1000 board feet. In 1929, timber sold for about $9 per 1000 board feet. In 1990, it was going for just over $500 per 1000 board feet. By 1992, some Japanese buyers were paying better than $1000 per 1000 board feet of prime timber.

(Some say the Japanese were importing the wood and sinking it to store it under water. Repeatedly getting wet and drying out are very hard on wood. Wood will last a long, long time if it is kept either completely wet, or totally dry. Perhaps some of those sinking the poles in Japanese waters were speculating that the price of timber would continue to go up and up.)

What would Alfred Nordgren and the other early twentieth century loggers have said if they were told wood would one time sell for better than a dollar for each foot?

In the 1920's, the Nordgrens had no way to know how lucrative owning timberland would some day become, but they understood they were much better off owning and working their land, rather than selling off any acreage. In fact, over the years, Alfred began piece by piece to buy up parcels around the Nordgren holdings, and Earl, over the decades, added more and more until he had amassed over 5000 acres of prime timberland.

But in the early days making a go of it on their own was no easier for the Nordgrens than hiring out to work for someone else. The only advantage was that they were working for their own future -- they hoped.

Alfred worked one spring and most of that summer cutting and hauling his logs to a saw mill near Gaston. The mill was having hard times, too, and they couldn't pay Alfred for the logs he brought them, though they promised again and again that they would as soon as they could. Finally, Alfred hired a lady attorney in Forest Grove to get his money for him. She was able to get half of it, but he never saw that money, because she took it and skipped town.

Alfred was coming to a conclusion about the hard facts of economic life. Cash and carry is the only way to deal. Promises do not fill your family's empty tummies.

Years later, Alfred sold the timber rights to a piece of his property to one of the major lumber firms of the Northwest. That meant that Alfred still owned the property, but the company had paid him for the right to come in with their equipment and cut the timber, haul it off and sell it. With other jobs in other locations, the firm did not get around to actually coming out to cut the Nordgren timber before the time limit in their contract ran out.

The managers approached Alfred, wanting him to agree to let them go ahead with the cutting now that they were ready, but

Alfred shook his head. He stuck to the letter of the contract and made them purchase the cutting rights a second time.

They were getting started setting up the operation when they realized they did not have a needed right-of-way across Alfred's property. Alfred shook his head again. He wouldn't give them a right-of-way. He made them buy that, too.

Alfred was teaching his sons to deal straight-forward. Do what you say you are going to do, and expect others to do the same.

Years later, Earl promised a saw mill in Scapoose that he would deliver a load of trees cut to their specifications and have them at the mill before Monday afternoon. The cutting took longer than Earl had estimated, so he had his crew work on Sunday to finish up. Just as the clocks were chiming twelve noon, Monday, Earl's trucks pulled into the sawmill in Scapoose.

The buyer scratched his head. "That is the first time anybody has ever done exactly what they'd told me they would," he said, extending his hand to shake Earl's.

A timber company's profits depend largely on the skill of their cruiser. "Cruising" timber means actually going out into the woods and assessing the value of the lumber you're going to get out of it. It means counting the trees and estimating their height and the number of board feet of lumber they will produce.

The accuracy of the cruiser's estimate determines how much the timber company will bid to buy the land or pay for the rights to harvest the timber off someone else's land. If the cruiser is right, and his company can make a good deal, they will make a profit. If he's wrong, all the loggers' work in the woods will profit them very little.

Archey Adams was out alone cruising one stand of timber. He marked some of the trees with a rag or with paint to be sure he wasn't counting the same trees twice. It was getting along toward sunset, but Archey was so close to completing his task that he kept at it, trying to get finished so he wouldn't have to come back out the next morning and lose all that time.

He finished, all right, but not before it had gotten so dark, he could no longer see his way out. With the darkness came bitter cold. With no extra clothing and no gear for setting up camp, he finally just ran around and around one of the trees all night to keep from freezing.

Art Vaandering was logging as a percentage contractor for Earl one time. Though the stand of timber had been cruised accurately and the contract was a fair one, the market price for the thousands of board feet went down after the deal was made.

Art was barely breaking even despite his hard work. He was selling the logs for about the same price he was paying Earl for the privilege of cutting them.

Art's wife was keeping the books for him. Sighing, she showed Earl the figures.

"You can't work for that," Earl told her. "You've got to make a living."

So he lowered the price Art was to pay him. Mrs. Vaandering looked at him in surprise and smiled. Art would do anything for Earl after that.

As with any estimating job, if the seller then decides to back out of the proposed deal, the cruiser has done the work, but there is no income.

A dentist once had Earl cruise his stand of timber out on the Cowlitz River. It took days to do the proper inventory. Earl finally had everything ready to make a bid on the timber rights, but the dentist changed his mind and decided not to have the trees cut. Those days of work were a dead loss for Earl.

On another potential sale on the Cowlitz River, Earl cruised and bid on a job, with the proviso that he could get the needed permits. That seemed fair to the sellers.

Earl checked with the Forestry Department and got permission to do the cutting. But then, when he checked with the county, he found out he would not be allowed to use their roads to haul the cut logs out to the saw mill. Earl told the sellers he would have to back out of the deal.

About a year later, a large timber company sent their cruisers out and made what they thought was a terrific deal. They had checked with the Forestry Department for permission to cut, but they did not think to talk with the county until after the deal had been signed and money paid to the landowners. The company's owner was so peeved about being hoodwinked that he defied them all and had his logs hauled out by helicopter. Of course, that is a very expensive way to haul. He didn't make any money on that operation, but at least he hadn't let them skunk him.

Alfred Nordgren had pounded into his sons the need to be straight-forward and calculating on any business deal -- or play the sucker. They were hard lessons for a teenager to be learning, but those were hard times. Even under those tough conditions, though, Alfhild and Fred and Earl found time for fun growing up.

Fred made himself a bicycle. It was not sleek, but it worked, and Fred was pretty proud of himself. But they lived up in the hills, and finding a level place to ride wasn't easy. One day he rode his handmade bike down the road and it got going too fast. To keep his balance, Fred cut a corner and ran right into a car coming up. Fred swerved to avoid hitting the car, and got into gravel at the edge of the road. He couldn't control the bike, and went for a terrible spill.

The driver stopped and hurried back, worried. "Are you all right?"

"Fine," Fred told him. His arm was scratched and bleeding, but he was too embarrassed to admit how much it hurt.

A few weeks later, the bicycle was stolen.

From high school on up into their manhood, Earl's best friend was the irrepressible Bill Franck.

"I really liked him," Earl says. "When we were kids he had a De Soto and I had the old Chevy. We were driving down to the beach. I was going 85 -- and he passed me.

"Once he was necking with his girl and not watching the road. He ran right off into some man's garage.

"Bill and I and a couple of girls went up in my Chevy into our old logging area. It was dusty -- you wouldn't believe. After a while we were coming back down that road miles an hour. Pretty soon, though, we had to slow down and that cloud of dust came down on us right through the back window. Everybody laughed to beat hell.

"One time Bill was driving his De Soto, fast, as always. He came hell-bent around a corner and right into a guy. The man jumped out of his car and stormed up, but Bill grinned and said, 'I don't know what you're so mad about. It was all my fault.' That really took the wind out of the guy's sails."

At the dances at Pumpkin Ridge, Bill Franck was a popular guy, always so much fun to be around. Once Bill left the dance early, and when Earl left about midnight, he saw a car in the ditch he was pretty sure was Bill's.

"I stopped and walked back. 'Are you all right?' I called.

"Bill rose up from where he'd been curled with his girlfriend. 'Yeah,' he grinned, 'we're making out real fine.'

"We were out driving in his car in high school and he had his girl along. We stopped at my dad's place and went in and introduced her. My dad says, 'She is pretty, but this isn't the same one you had along last time.' You should have seen the look between those two.

"Bill never paid attention -- never saw the consequences of what he'd do. He had his wife Franque's grandmother's piano on the back of his pickup once. It was a real antique as well as a family treasure. But Bill came roaring toward Forest Grove and turned a sharp corner in Watts. He whipped that piano right out the back of the truck. It landed on its keyboard upside down on the Gales Creek Road, smashed to smithereens. It was tough telling Franque.

"Bill and Franque were so spontaneous. What one didn't think of, the other did, and each was always ready to go along with the other. They got into more scrapes that they could easily have avoided. One day Bill came home and said, 'Let's go to Yellowstone.' So they piled into their car and took off for Yellowstone. When they did come back, their deep freeze had burned up."

Bill learned to fly. Franque went up with him once, but he did such crazy things even she'd never go up with him again.

Bill was blasting stumps for Walt Hayden. Walt bought logs. He was probably worth a million dollars then. But his eyes were so bad he wore thick, heavy lenses in his glasses. Walt went up to help Bill light the fuses for the stump blasting, but his vision was so bad, he was setting fire to fern stems. Bill followed him around, but it wasn't easy trying to light the fuses instead of the ferns without making Walt feel bad.

Walt "helped" as long as he could, then had to leave to get to a business appointment in town. He was hurrying away when Bill got up the nerve to ask him for a loan to start a logging operation down on the coast.

Walt was in a terrible hurry, but he stopped and got out his check book. "I don't have time to fill this in," he said as he signed his name and gave the blank check to Bill. "You just put in what you need." Bill was pretty proud of that.

The Francks liked to hunt. Bill shot a deer once up in Carpenter Creek and dragged it over and put it in the trunk of his car. But his shot to the head had only stunned, not killed, the deer. As he drove home, it revived and started kicking and thrashing, making a terrible racket. Once he got home, Bill opened the trunk and jumped back. The deer's legs were slicing and flashing out at him, but somehow he got a hold on them until his brother Bob could finish off the animal.

Bob Franck was a lot bigger than Bill. Bob loved to ride him, sassing him real bad sometimes. Finally Bill exploded and hit him so suddenly and so hard that Bob was out cold for five minutes. Bob didn't sass him much after that.

But all the Franck boys could sass and rag. Bill got into to a mood one time with his other brother Dick. But his timing was not good. The two were out alone logging deep in the woods.

Dick was working the lines with the donkey engine, using its steam to power the huge cables lifting and moving the cut logs.

Bill, without calkshoes or climbing rope, had strapped into the harness to be hoisted on the strawline up near the top of the spar tree. The Francks had finished cutting and moving all the timber that could be served by the haulback and bull blocks on that side of the spar tree. Bill was up there loosening the straps so the heavy blocks could be shifted to the other side of the tree at the same height. From that side, the cables would be clear to lift logs cut from the stand on the other side of the spar tree. They couldn't log full circle around the spar tree without shifting the blocks, or they'd foul up the lines.

It had been a long, rough day in the woods. Dick's inattentive, almost careless, working of the donkey engine had lifted and jerked Bill on the line again and again. Tired, uncomfortable, and repeatedly scared, Bill was getting testier and testier until finally he cussed his brother out.

Dick, on the ground, kept working the lines for a while, but he was getting pretty tired of his brother's mouth. When it got about as nasty as he would put up with, Dick just tied off the friction line and headed home, leaving Bill stranded up there in the top of that tree with no way down.

It was two hours before a forestry agent happened to walk that way and heard Bill hollering. Dick stayed out of Bill's way for a day or two after that.

Earl first worked as a high climber when he was eighteen years old. One of the old climbers gave him some tips on what to do and how to do it. Earl listened closely, then put on his calk shoes, strapped on his leather belt with all the heavy tools, and started up the tree.

His first spar tree wasn't too high, only about a hundred feet or so. Earl cut the branches and made it all the way to the top all right. He set up the blocks and pulled through the lines. Tired, but inwardly grinning at what he'd just accomplished, Earl signalled that he would come down on the line rather than climb down on his own.

He was pretty sure the men would find some way to tease him in honor of his first job of tree topping. He was pretty sure it was only a joke when they slipped the brake on the donkey engine and he started, not to be let down smoothly and easily, but in near free-fall. They let him plunge sixty of those 100 feet before they caught him up again and lowered him slowly to the ground.

Earl saw the men laughing as they stood around the base of the new spar tree, watching him. He grinned, once, unhooked and walked away as steadily as his knees would carry him.

"Keep your chin up," he whispered to himself. "It helps keep your mouth shut."

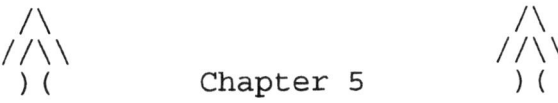

Chapter 5

"you can only get so scared"

Anyone who has split wood into fire logs or who has accidentally pounded his own thumb instead of the intended nail can begin to appreciate what a tree topper is up against.

He is working with only his short spikes grating against rough tree bark to keep from sliding down 20 or 40 or 100 feet. His weight is -- counter-intuitively -- leaning out away from the tree into a slender, but strong, rope, which is also held up only by its friction against the bark on the opposite side of the tree. And then he is supposed to be able to work with axe and saw.

He is swinging with a sharpened blade against a tree on which he is balanced. That tree shimmies and shivers and sways in reaction. He's trying to hit a moving target.

He is also clearing away the limbs as he moves up the tree. If he were to fall, he has already removed whatever there had been of a "safety net" of branches behind him that might have helped break his fall.

"When you climb to the top of a 150' tree, it doesn't look so high from ground level," Earl grimaces. "But looking down from the top of that tree, it looks like it's miles to the ground. And then the damn tree sways.

"You can't work at all at first, but then you just do. You can only get so scared."

When the cut just above the topper is completed and the top of the tree falls away, its final splitting off sets the whole rest of the tree trunk swaying in reaction -- and the topper, too. Sometimes he is swung out ten or fifteen feet as the upper end of the tree bends. And then he is whipped back with it as it straightens. But the wild ride isn't over yet. He is swung out six or eight feet again as the tree bends the other way.

Sometimes the high climber's gear gets swung out on its own momentum so its dangling rope extends out farther than the topper. Loggers say it is an interesting experience to watch your axe change direction and come slicing the air directly back at you as you are whipped uncontrollably at 120 feet above the ground.

Usually, the tree sways again, and the topper is yanked out of the axe's path.

When the swing and sway finally come to a halt, the high climber has to start the real work of setting up the pulley blocks. He has dangled a pass line from his belt since starting up from the bottom. A crewman on the ground attaches to it whatever equipment is needed, and the high climber pulls it up by arm strength.

But the pulley blocks, which are made of iron, are often more than four feet square and two feet in depth. They are far too heavy for any man -- or several men at the same time -- to budge even when they are on the ground. No man can haul one up 100+ feet as dead weight.

So the topper begins by hauling up cable bands that he cinches tight around the tree high near its top. The topper encircles the tree with the cable band and secures the block with a clevis to fasten it tight.

Sometimes the bull block is secured by pounding small spikes into the tree through holes built into its hooked metal plate. But usually even the bullblock can be cinched tight enough to hold without nailing it into the tree because the trunk itself tapers. The weight shackled onto the metal plate digs the plate deeper into the wood at its lower edge.

First, the small strawline block is shackled onto the spar tree up near its top. A rope is fed through the pulley and allowed to dangle back down to the ground crew. They attach one end to the strawline drum of the donkey engine. To the other end, the ground crew ties the first of a series of thicker and stronger cables, which attach to each other with a built-in hook and eye for a smooth seam to keep from jamming the pulley.

The donkey's steam (or, in the Nordgrens' case, gas) boiler powers the heavier cable up and through the strawline block until it is of sufficient strength to be able to haul up the two, heavier, haulback and bull blocks.

That first, comparatively light, strawline block will stay fixed to the spar tree to act as the pulley for the strawline or pass line, now strong enough to be able to lift other climbers high into the tree. A "harness" of two chains encircles the climber's legs and lifts him safely. Having the donkey engine pull them up to the needed level saves the men the time and labor, and danger, of climbing up each time with a climbing rope.

The pulleys high in the spar tree have to be greased as they are set up -- and frequently as they work -- to keep them turning smoothly. Roughhouse Dickson used to climb up to the top of the spar tree to grease the block, then clean the grease off his hands by wiping his fingers through his hair. It wasn't a fashion that caught on, however.

The next block hoisted by the strawline is the haulback block, and then the massive bull block. Both are huge and heavy, made of iron. The bull block houses the pulley that lifts the cut logs high enough to clear obstacles on the ground so they can be moved to where they can be transported to the sawmill by truck or railroad. Strong cables threaded through the bull block are allowed to dangle back to the ground, where the crew attaches one end to the bull block's drum in the donkey engine. To the other end -- after stringing through the complicated system of sky lines -- they attach the butt rigging, which includes the choker to secure the logs to be lifted.

Once the log has been moved to where it is wanted on the ground, the whole long, weighty cable must be hauled back out into the woods so the hooker can set the choke on the next log to be moved. That is the job of the haulback block on the spar tree. The haulback line is attached to its own drum in the donkey engine.

There are also one or more haulback blocks on ground level in the woods, attached to stumps. The cable is threaded through each in turn in an elaborate system of ground lines that can be perilous to the unwary. Even the men on the ground who know where they are and how powerful their movement can be, have been caught unaware when a dangling line is pulled taut.

Without tension, the line will have a sag or bow as gravity pulls on its unsupported middle. That bow is called a "wow." When the donkey engine begins its pull, the tension in the line is transferred through the whole cable system, and the wow may straighten suddenly. Experienced loggers know to give the line full clearance, but a young logger or water boy might walk too near a still line, not realizing the danger.

The entire cable system is huge, massive and complex -- until the final lifting motion is supplied by the bull block on the spar tree. Four, five, even six guy wires are needed to counteract the huge forces on the spar tree.

Because of the steep layout of the land they were working, one logging outfit needed an unusually tall spar tree. They spliced one spar tree on top of another one. High climber Ernie Lyda said there were so many guy wires from all directions holding that double spar tree upright that you couldn't shoot a .22 up in there without hitting something.

Ernie Lyda was the little bit of a guy famous even among high climbers for his daring. He'd top a tree, hold on until it stopped its sway, then clamber up and stand erect on the sawed- off top.

He'd shimmy back down into position on his rope and drop his cap. Then flipping his rope and scrambling, all but leaping ten or twenty feet at a time, he got so he could beat that cap the hundred-plus feet to the ground nearly every time.

Earl Nordgren never did that standing-on-the-newly-cut-top trick. "I have the best survival technique of all," he laughs. "I'm a coward."

With the spar tree and donkey engine as the center, the loggers cut and cleared, working in a half circle. Then, after resetting the bull and haulback blocks to the other side of the spar tree, they worked the other half circle. Of the original square of timber land to be worked, it was difficult and costly to reach into the four corners. Most often those corners were merely left as virgin timber.

In the Great Depression, small, shoestring "gyppo" logging outfits would come in with caterpillar tractors and log the four corners left by the larger companies.

The virgin timber left might well contain a number of peelers, which are huge, thick-trunked trees that are spun under the knives of the sawmill. Their wood just peeled off in curled sheets the length of the log. There was money to be made from those peelers, but usually not enough so the big outfits could afford to set up separately to cut the four corners.

Pine grows mostly at higher altitudes. It is very soft wood. At lower altitudes, fir grows best. Fir is soft, too, but not as soft as pine. But fir has a lot of pitch. The loggers have to oil their power saw chains again and again to keep the pitch from gumming them up. Large, old fir trees might all but "bleed" sap when they are cut.

Tall, straight pine or fir trees that can be used for poles command a good price, providing they can be harvested without breaking their slender 120' or 150' lengths. No matter how carefully they are cut, many split or break as they land. Intact, they are worth so much that Earl devised all sorts of ways to save them from splitting. He even tried using a crane to lower them to the ground, but that was too dangerous.

Finally he hit on a simple, but ingenious solution. If the pole were cut so it would fall onto a mound of dirt and brush and leaves, the mound would cushion the blow. The log would land, rocking freely up and down on the mound, scattering the energy of the fall. The technique worked so well that if there was no natural mound handy, Earl used the cat to bulldoze one where he needed it. Slick!

A school marm can be a special problem for a topper. A school marm is a tree that starts growing as one trunk, and then, because of a lightening strike or other severe event, forks out as two stout trunks growing side by side, sharing a single base. That second trunk can foul the lines as logs are swung around. The high climber has to choose one trunk and cut down the other just above the fork, however high that may be.

The weather on the ground may be calm, but above eighty feet in the air, there is almost always a wind. When a topper works in a wind, he avoids undercutting for fear the wind may shift direction. If he's cut too deeply into the tree at one point, the shifting wind could push over the whole top. Because it had only been cut partially through at that point, the tree could slab out, splitting apart with enormous force, yanking apart the high climber's life line and hurling him to the ground.

As Earl puts it, "That could ruin his whole day."

So if there is any appreciable wind, a topper will chop around and around the trunk instead of deeply at one spot. That way, no matter which way the top falls, it won't split out the tree and take his rope -- and him -- with it.

While the spar tree is being readied, the ground crew brings in the donkey engine and sets it up near the base of the spar tree. With its powerful steam or gas engine, the donkey is made to haul itself into position across the uneven forest floor. The main line is wound around a sturdy stump or tree on line in the direction they want the donkey engine to go. The donkey then draws on that line, hauling itself on skids.

Jim Burns was working the donkey engine while several men were rigging a new spar tree, putting up the blocks and lines. Three toppers were working and five buckos, each with a line.

It was foggy that morning so Jim at the base couldn't really see Ed Medwood or the men above him. Jim got the signal -- one jerk of the line -- and knew Ed wanted to come down, fully expecting to be lowered reasonably slowly. Jim worked the friction to bring him down the pass line, but he somehow got it wrong. Jim knew, though he couldn't see him, that Ed was coming at a dead drop.

"Oh, boy!" Jim whispered and flipped the friction.

Ed had already fallen so far that Jim could see his feet. Then the change in the friction sent Ed back up as fast as he'd been coming down.

"The next time Ed wanted to come down, I knew better than to let him touch ground," Jim laughed, sheepishly. "I lowered him to within four feet of mother earth, then tied off the line and took off running. I never did go back that day at all."

A single operation in the woods is known as a "side." A side consists of the spar tree; the yarder donkey (donkey engine) and its operator, the donkey puncher; the "hooker," or choke setter, who secures the choke and supervises the moving of the logs; plus all the cutters and men working to move the logs.

At times a side will include even a sawmill and its pond.

Besides the log pond, the timber company would build a reservoir, usually from a spring. They'd pipe the water to somewhere near where the donkey engine and the railroad steam engine were working.

A twelve year old boy's first job in the woods might be as water boy. It was up to him to keep the water tanks filled for the steam engines.

A second job as the boy grew older was often to act as "whistle punk." Jim Burns started in the woods at fourteen years old as a whistle punk. It was his duty to keep the men informed of what was happening by signalling to the donkey puncher what the hooker told him he intended to do.

A wire was stretched between the donkey and the whistle punk, who stood near the hooker. The wire went over the young punk's shoulder. When the hooker yelled, the punk pulled on the wire, and the steam whistle on the donkey let out a piercing yip that could be heard for miles.

The pattern of the whistles told the donkey operator and the rest of the crew that the hooker wanted to lift or move the log forward or backward. A single yip meant, "go ahead." Two whistles meant, "back up." Four meant, "slack the main line." Of course, whenever anything went wrong, it was always the whistle punk's fault.

Jerry Kemper remembers a greenhorn boy working as whistle punk. Not knowing any better, he walked under the guy lines and work lines on the spar tree. As a log was hauled, the cable split off a large chunk of wood that crashed down on the boy. The men dragged him out, limp and frothing at the mouth.

"He was like a rag," Jerry remembers, and even forty years later, his face stiffens and goes white as he tells the story. "We called for a stretcher. They did what they could, but the boy died."

The men had known when they saw him that he would. Danger of maiming or death is a reality for those working in logging.

It takes constant alertness to avoid danger.

Bill Franck was salvaging with his crew in the Tillamook Burn area when he spotted a butt end that could let go and fall on someone. Bill stepped out of sight to get at the butt end. But it whipped and pinned him against a stump. Bill was hurt bad, but no one could see him and, with their power saws roaring, no one could hear him.

It was quite a while before one of the men turned off his chain saw and suddenly heard Bill's cries for help. It took four hours just to get Bill out of the burn area to an ambulance. He was in the hospital a couple of months that time.

Logging before, and even after, the introduction of the steam donkey has always been hard, physical labor -- and dangerous. Perhaps the only comparable job in terms of exertion and danger today is that done by

Mohawk Indians who work the steel girders in erecting high rise buildings.

Old-timer loggers will tell you that early in the twentieth century, some of the big timber companies averaged a man killed every day they worked, in one or another of their operations. It got so common, the boss would simply have the body dragged behind a "cold deck," or stack of cut logs, so it would be out of sight until the end of the work day. It would then be loaded behind the men onto the back of the beat-up, little "crummy" bus that took the exhausted loggers back to camp.

Loggers early in the twentieth century were rough men, or men who grew rough to survive the harsh conditions. Yet the economy in the Northwest was such that nearly everyone in the area west of Portland logged or had to do with logging to make a living, especially during the Great Depression. Many of the early loggers had never had the luxury of much education. They were men just trying to earn enough to keep themselves and their families alive.

A logger's wardrobe may have consisted of long-johns, rough pants and a "hickory shirt," made of a tough material striped in gray and black. Dressing up for an occasion meant washing his one set of clothes. An old man dressed in his new-washed hickory shirt and rough pants rose up, hat in hand, at the funeral service for a fifty year old who had been killed in the woods.

"He was a young fellow," the old man droned, "raised lots of cows and kids and he thought, 'By God, they'll take care of me in my old age.' Only he never got there."

Most of the time it was each man making his own way and looking out for himself. There was no welfare, and few men had insurance. If he was hurt or killed, his family was most often on its own.

But men had to have work, despite the danger and the conditions. Loggers were out in the elements, always. In summer, there was dust and heat and insects, and even wild animals. More than one logger has had a run-in with a bear. And there is the unrelenting danger of fire.

Winter logging has its own set of unpleasantries. The tractors with rubber tires, called skidders, often did not cope well with winter conditions. But even the metal-treaded cats could have problems as the mud in their tracks froze solid as cement.

If they were near a sawmill, the cat drivers would drive up onto a sawdust pile. The warmth from that sawdust would keep the track mud from freezing over night so they could roll in the morning. Otherwise, the drivers found it far too easy to strip the gears trying to get the cat to move.

Men who logged all winter often found themselves working standing, kneeling, even lying in the snow - or in ice water. One day the weather was so rough, Harrison Heisler worked all day long and only got out one log -- and that he had to use to fuel his steam donkey.

During the winter, Earl had to find equipment to rock the roads and make them more nearly passable in the rain, and mud, and ice. Swiss cat driver Joe Zberg was trying to get the diesel cat up a steep hill covered with ice. He got it nearly to the top, then ran out of traction. The enormous caterpillar tractor started sliding back down. Joe Zberg jumped, or was thrown, clear, but the cat rolled -- and ended upright on its treads.

Rain is a near constant in Oregon winters. Cold mud is a common milieu.

Sometimes the cats were tracking in glop. But in the winters of the early 1970's, the Nordgren crew was getting out 90 and 92 foot poles to send to Guam, and it was worth the effort.

Earl sang his own version of "Love Letters in the Sand":

"Oh, it was a day like today,

we passed the time away,

writing love letters in the mud." He grinned, then, and explained, "He was an Oregonian, you see."

On a winter job off Roderick Road, Cecil Thrush fell off a log into mud up to his waist. He was in so deep, he couldn't move, and no one else could get near him. Earl had to pluck him out with the tongs.

Chapter 6

"Any sass, hit him on the ass"

Old-timers consider today's logs merely "brush." In the early days, they wouldn't have thought it was worth logging when there was so much bigger, heavier timber everywhere to be cut.

Jim Burns, born in 1909, remembered standing and looking out over thousands of acres of timber in the late 1920's.

"You couldn't have told me in those days that anybody would ever run out of timber to cut."

Timber in the late 1920's went for about $9 per 1000 board feet. In 1990, that same timber would go for $500 or more per 1000 board feet. But the difference in price is not simply a matter of money.

There is a terrible difference in the pace of logging. It started getting faster and faster when the steam donkey and cat logging came in. There was always a crew coming, a crew working, and a crew going home. International Paper kept seventeen "sides" working at one time. Each side had to produce 100,000 feet of timber every day.

Dave Britton was a hooker, or choke setter, who worked primarily for I.P.. He was a hard-nosed boss under whom every man had to produce or he'd be run right out of the woods.

When Alfred Nordgren started logging his woods, the cut logs were moved by teams of horses. The logs were secured with what look like gigantic sugar tongs. The hooker set the ends of these huge wishbone-shaped iron tongs by digging the nasty spike at the end of each arm into the log. A ring forged in the top of tongs was used to attach them to the horses' rigging. The team of two, four, six or more horses then pulled the felled logs by main strength into position to be hauled away by truck or railroad.

Earl remembers a brief period of regulations that required horses to be diapered while working in the woods in a park. That particular regulation did little to further the loggers' respect for government rules.

With the introduction of the donkey engine, instead of tongs, a "choker" was used to secure the end of the log to be moved. The hooker was now called a "choke-setter." He "lassoed" the noose-like cable around the end of the cut log and cinched it tight with a sliding metal "bell."

The choke was the working end of the butt rigging that was attached to the main line up through the bull block, perhaps by way of a "sky line" that was strung out away from the spar tree. The horizontal sky lines gave the choke-setter the ability to move the logs some distance above the rough terrain. The bull block housed the pulley through which the main line was strung from its drum on the donkey engine.

Jerry Kemper worked as a hooker, the man who chained the logs to be hauled to the truck or for the crane to lift and load. Jerry loved his work, and the responsibility that went with it.

"I felt free working in the woods," Jerry said. "I was good at it, and I could get a job anywhere I wanted."

One time Jerry was maneuvering the butt rigging. Butt rigging is always a challenge to haul around, but this time there was the added danger that it had to be moved over rimrock. They were just making headway, when the whole rimrock gave way. Tons of earth and rock shifted and tumbled down at him. Jerry ran literally for his life and just jumped out of the way.

Jerry's cousin, who had seen this happen, threw up his hands and walked away from logging altogether.

"That's it!" was all he would say. He never did come back to the woods.

Even in the best conditions and with modern equipment, logging is demanding work. In the early days, so much more had to be done by sheer physical exertion.

Sometimes the men worked so desperately hard in the woods that, when the whistle blew to end the day's logging, they would lie down beside the tree they'd been working on and sleep for half an hour before they could walk home, or to the camp set up by the logging company.

In the first half of the twentieth century, Weyerhauser, International Paper, Spalding, and Gales Creek Lumber all had camps in the woods west of Portland.

In the camps, the shelters for the men were crudely built shacks with new straw for mattresses every Saturday. The men slept in their underwear -- heavy black long johns with white buttons -- while their clothes dried by the woodstove overnight. Even wet, their underwear was warm.

One of the old loggers who worked for Alfred Nordgren in the early days wore the same sewn-up pair of long johns all year round. He would bathe once a year, unless he fell in the creek. That was extra.

It was a rough life for the men. It was incredibly rough for the few women and children who joined them in the larger camps. Early in the twentieth century, children of the Northwest could not always go to school. Many were raw-boned and rough-edged because they were needed at very young ages to help work the family farm or, as they grew stronger and more agile, to work in the woods.

Martha Norris married a logger in those early years. She and her husband moved into the back woods logging camps, where she found the way of life even harder than she had imagined it would be.

A family friend in Wyoming had built a sturdy cabin for his bride. When he took her to see her new home, she was thrilled -- except that there were no closets and they had no chifforobe.

"I don't have a suit, and you don't have many clothes," he told her, pounding a nail into the wall. "There, that ought to take care of us."

For Martha, even her own bare wall would have been a luxury. In the logging camps, many times she and her husband scarcely had the bare necessities. Privacy was something they got only if others accorded it to them when they drew a blanket across to divide the cabin's single room.

Women's work was never done. But it was often tackled as a team by the few women who stayed with their men. Cooking was done in giant black, wrought-iron kettles. The men, starved and weary as they arrived back at base camp on the railroad or tramping on their own tired feet, ate enormous portions, then rolled into their cots in their long johns and slept the sleep of the dead.

Martha Norris worked hard to fit in with this rugged life. She couldn't bear watching the camp children growing up without some sort of education. She had had some schooling herself, so she gathered them together to teach them what she could.

If the logging job involved a large acreage, the loggers and their families might live in one area or several close camps for months at a time. Then there would be time and a sufficient number of children to make nearly a regular school situation, at least until the older boys had to go off into the woods to help with the logging.

Martha remembered one of the older youngsters giving her a particularly hard time. She warned him, and when he continued acting with disrespect, she slapped him hard across the mouth.

The next day she got a note: "Don't never smack Henry across the mouth again. If he gives you any sass, swat him on the ass."

Martha was a survivor. She was direct because she had to be, but her humor made her a joy. When you were near her, her wry outlook on life made you laugh until there were tears in your eyes and a hitch in your gut.

But when she was driving, she never waved back at friends on the side of the road. She never recognized them. It was likely she never even saw them. Behind the wheel of her big, old Dodge, she never looked right or left. The only traffic ticket she ever got was on the two-lane road from Forest Grove through the Coastal Range to the ocean-side community of Tillamook. She was cited for going too slowly, holding up traffic for miles behind her.

Driving too slowly was never one of Earl's problems with cars. Earl had a brand new Nash about the first year that windshield washers came out. Art Vaandering had told young Earl that his new sedan "looks like a bitch dog that's just been fixed," but Tom Prince of Hayfork, California, was more impressed.

Earl was driving Tom Prince out of Hayfork on a bright, clear, sunny day when he decided to clean some of the bugs off the windshield. He pressed the lever for the washer. Tom leaned forward and cocked his head to look up.

"By gosh," he said with wonderment, "I think it's raining." He'd never seen such a gadget and could hardly believe Earl when he showed him how it worked.

Tom Prince was one of the loggers who worked with Earl in Hayfork. The two men were walking with a third logger, who admitted that he had eight kids.

"Lord," exclaimed Tom, "I think you're going to seed!"

Earl worked cutting and hauling logs that long, cold winter in Hayfork, and it was like going back in time to the days of the early 20th century.

He stayed in a small, shake cabin that was merely a single room. It had a stove for heating and cooking, but there was no running water or electricity.

There were no towns near. Loggers and others from miles around would come to Hayfork on Saturday nights to visit the tavern. When asked if most loggers were heavy drinkers, Earl answered, only partly tongue-in-cheek, "Most of them had to be -- they needed that antifreeze to get them through the winter."

A small group of men who had been enjoying an evening at the tavern decided they would race to the next town. They all piled into their car to start off down the single narrow, winding country road. But they never made the first turn.

Hearing the crash, people went running out to help them tumble out of the wrecked car. Someone asked them if they'd been drinking. The driver drew himself erect and protested in a pained-by-the-implication voice, "Never touch the stuff."

The tavern in Hayfork was rough and tumble, but it was also the only place to buy a meal. Earl was there one night eating dinner -- a big, juicy steak -- when a man rode in on his horse. He rode right through the saloon and then turned in his saddle to shoot out the lights. Earl grabbed up his plate and ducked under the table to miss getting hit by flying glass.

Besides logging, gold mining was a large part of the Hayfork community income. One gold miner celebrated each new gold find by buying himself THE status symbol of the day, a brand new Packard automobile. He'd keep that car until it no longer ran, or until he found new gold. Then he'd run the old Packard into a shed behind his cabin. He had three in there.

He disliked the color of one of his Packards, so he got a can of paint he liked better and re-painted his prized car with a brush. It made a unique effect. Nobody could steal it. Everybody who saw it knew instantly who it belonged to.

People are curious, especially in small communities. Occasionally someone came up into the woods to watch the loggers work. One farm couple hiked up to ask to watch the trees being felled. It was near lunch break, so Earl invited them to come back up after the men had eaten.

Smiling their appreciation, they walked away as the men sat down on logs and stumps and brought out their brown bag lunches. As they finished, one young logger moved up into the trees. Earl, realizing where he'd gone and why, started speaking in a loud voice as though the old couple was returning through the woods beside him.

Gasping, the young logger came crashing down through the brush to get out of sight. Running, he couldn't get his pants up and, with them down around his ankles, he could barely keep his balance on the uneven ground.

It took him a couple minutes to realize that the couple had not yet returned, and that he had been royally had. Red in the face, he vowed revenge, but his growls were drowned in the laughter of the whole crew.

They did keep an eye out for revenge, though.

Loggers work hard. They have little patience with people they see as being unwilling to work. Many loggers had strong, negative reactions to long-haired "hippies," just on appearances. A logger describing a disreputable type might say he was the one in the "snagged-off pants."

One long-haired young man stepped into a beer joint in Gaston. The loggers inside took one look and swarmed to the door to toss him out.

Another long-haired young man found Earl working in the woods and asked him for a job.

Earl stopped and stared at the tangled hair draped around the young man's shoulders and told him, "You ugly devil, I wouldn't give you a job if you paid me."

The kid rocked back on his heels, then opened his mouth before stomping away. "You don't look so hot yourself," he retorted, but he didn't get the job.

Some of the loggers had even stronger feelings against hippies. A crummy, hauling a crew of weary loggers home after a hard day's work, overtook a young man with hair to his waist, who was walking along the logging road. The hippie lifted his thumb in the universal gesture of a hitch-hiker asking for a ride. As the little bus stopped, the hippie, grinning, ran to catch up and climb in.

But the loggers yanked him down in the back of the bus and held him kicking and screaming while one of the men sheared off several feet of that long hair with a power saw.

The hippie got the last word on that one. He sued and was awarded $2500. But he never again tried to hitch a ride on a crummy.

Chapter 7

Hauling Off

One old-timer resisted the new ways. He didn't like machines. He wouldn't use a steam donkey. He wouldn't even use a tractor. He did all his work, even his plowing, with horses. When he died, they pulled his coffin in a horse-drawn carriage.

While mechanical devices helped ease the horrendous physical labor involved in logging, at times the machinery itself caused problems.

One choke-setter went through any number of tin hats. Whenever anything went wrong, which was pretty often in the woods, he'd throw his hat down and stomp on it.

Jerry Kemper was another choker setting up a brand new sky line. The line was strung across a canyon and the crew was securing the "tail" end of the inch and a half thick cable around a huge snag which was six foot through, but ten feet tall.

Ordinarily the tail hold would be wrapped three or more times around a stump and secured with spikes to be sure the end would stay no matter how much pressure was put on it. But this snag was too tall to simply throw the second loop up and over it again once the donkey had pulled the first wrap around it. So the men secured the first wrap with 60 to 70 spikes.

Hoping that first loop and spikes would hold, they detached the end of the line from the donkey and were physically pulling the greasy cable to circle the base of the snag to create the second wrap, intending to secure it with spikes, also. And then they would drag the line once more around the base of the snag and secure the third wrap as well.

But the weight of the skyline hanging across the canyon was simply too much for the single wrap to hold until they could get the second and third wraps in place. Once the line started to slip out of its wrap, the weight and tension of the whole line across the canyon drew it so fast, the man had no time to jump out of the way.

Jerry Kemper was in the middle of the trailing loops of cable. As the end of the line whipped, Jerry was hurled into the air, up and out of sight.

He landed on his feet in a clump of low spreading vine maple trees that broke his fall. But his brother and the other men he'd been working with couldn't see him. They only knew that the way he'd been flipped, there was little hope for him.

When he got up and walked back to them, the others were white as sheets.

"What's the matter with you guys?" Jerry grinned, trying to make light of what could have been a tragic situation. He knew they hadn't expected him to be alive, let alone be able to walk out on his own.

In the early days, the hooker or choke setter would set the "dog", huge iron tongs with their sharp ends poked into each side of the log. The tong ends were joined in the middle, and at the front end was an iron ring where the hooker drew a chain through and attached it to the horses' rigging.

Eventually, with much sweat and a few cusses, the log would be ready to move. The horses would haul each log across the forest floor to the loading area, which was a bank or other natural (or constructed) elevation so it would be level with the bed of the railroad car or truck parked on the low road alongside the mound. The horses were then unhitched, and men would, hooking and dragging, muscle the logs onto the truck or car-bed.

Before loading cranes were built, men had to load the cut logs onto the back of the trucks by hand using "peevies." A peevy is simply a 5 or 6 foot pole about four inches thick at one end with a point and a sharp iron hook. The pole tapers down toward the other end to easy grasping size, maybe two inches across. The peevy is used like a lever. The pointed end is used to push and poke and prod, and the hook to grab into the bark and tug with.

Just down the logging road or farther along the railway line, taller "rollaway" loading platforms had been built, so the men could stack more logs onto the pile by moving them across at level. The higher the mound, the taller the load that could be added onto without the men having to try to lift the logs. It was difficult enough just to muscle them across on the level.

Moving the cut logs was heavy, tricky work. A unbalanced log could slip and wedge or break or crush an arm or foot or leg -- or a man.

Even today with cranes to hoist and pivot and stack the logs, the men on the truck depend for their safety on the skill of the crane operator as they push and shift and help balance the load.

The loggers used their axes to knock pegs or small branch stumps out of the tree trunk so the log would roll.

Most early logging truck drivers carried an extra Ford axle with them to help pry the logs. Henry Ford was reported to have gone as far as South America to get the strongest steel he could find to make his axles. The loggers prized them because they were more durable than anything else they could use.

Fully loaded, the truck hauled its logs away from the logging site, out of the woods to the saw mill where the driver dumped his load into the mill pond.

    Logging roads were often not much more than swaths through the brush cleared by a bulldozer. If they were to be used for long or in bad weather, they might be "rocked," or covered with gravel.

    Old Bud Vanaken used to haul rock for the logging roads. He'd get growly when his false teeth hurt. "These teeth," he'd moan. "They fit better in the box than they do in my mouth."

    A hillside logging road is usually created from the top of the hill, down. The caterpillar tractor is maneuvered to the highest point of the logging site. The tractor driver then zigzags down the hill, keeping a regular grade, bulldozing as he goes. More often than not he will come out about where he wanted to. But sometimes the terrain is so irregular that another man may walk ahead of the cat, blazing the trees by cutting marks, painting swatches or hanging small red flags. The blazed trees tell the cat driver which way he is to go, but it still up to him to judge the grade and clear a drivable rough road.

Howard Wahl was working a cat higher up the slope than most of the rest Earl's crew when he tried to move a long log that looked innocent enough. What he did not realize was that the log had been a bee tree, and the bees were already angry.

When Howard disturbed that tree again, the bees were furious. They swarmed to the attack, stinging Howard everywhere there was an opening, even getting under his hard hat.

Howard leaped off the cat and ran down the slope all the way to base camp where the men caught him and helped swat away the remaining stubborn bees.

Earl ran up the slope to get the cat. Howard had left the engine running. Understandably.

Another cat driver was working near the Pacific coast digging out the side of a mountain, when suddenly he was caught in a landslide. The whole mountain seemed to be caving in on him. The cat was buried completely, with the driver inside.

Caterpillar tractor cabs are built strong enough so that even if the whole tractor rolls over, the driver is still protected. So the driver was unhurt, but he was buried alive! With no one near enough to know yet what had happened to him, he knew he had to dig his own way out.

He made it, but he swore he'd never drive a cat again.

"That was too close," he whistled.

Earl had a young fellow driving truck for him. The kid was game but not very experienced. It started to rain hard while the crew was loading the kid's truck on the landing. They were logging far up a mountainside, and the loading platform was three or four hundred feet up a steep slope. The logging road was merely a clay scar without much gravel and no grading.

When the kid's truck was loaded, he started away from that landing platform much too fast, evidently not realizing that the rain would make that clay as slippery as a sheet of ice. Earl and the crew saw his brake lights go on, but by that time it was far too late. The truck went skidding straight down.

The whole crew took off running after him. By the time they caught up with him the truck had rolled to a stop on the flat. They dashed to the driver's side to see if he was okay. The kid leaned out his window, his face white.

"That goes kinda fast up there at first, don't it?" he said, trying to grin.

Earl sighed as we were on the highway passing an 18-wheeler with extended cab for living quarters. Even on that hot day, the truck driver had his windows rolled up.

"He must have air conditioning," Earl sighed again. "They've got such nice cabs nowadays. Nothing like we had in the early days."

Driving a log truck is no picnic, even now when it's got power steering and power brakes. And air in the tires. Alfred Nordgren's 1919 log truck had hard tires when there were a few ruts in the road. Actually, there was a little road now and again around the ruts.

When not loaded, modern log trucks carry their own trailer lifted by a built-in hoist. The long bar that sticks up from the trailer at a slant over the cab like artillery is called a "reach."

When loaded, the logs rest on swiveling "bunks" on the truck and trailer.

Side bars keep them from rolling off. The logs themselves provide the connection between truck and trailer, but, even with their swiveling bases, would cut corners at every turn if not for the long reach hitched beneath them into the back of the truck. The reach helps extend the front of the load into the outer arc so the trailer better follows the truck around the curve.

Old Meek was a log truck driver who tempted fate now and again. His truck started to roll once when he'd stopped on a grade to check his load. Meek ducked under the reach to get into the cab and stop it. Those who watched held their breath. He could easily have been crushed.

Turning corners with a load that long can be quite an art, but Meek seldom used his mirrors. At one curve, he flipped his trailer and load of logs.

Earl, coming up behind him, berated him angrily: "You don't even know where your trailer is!"

"Oh, yes, I do," Meek responded. "It's in that field down there."

Even using the mirrors, making turns, especially in town, can be a problem with stiff long loads that do not bend in the middle. Earl was driving a load of 120 foot poles up the country end of south 'B' Street into Forest Grove. He had to make a right turn onto 19th Street. 19th, a one-way heading east, has two driving lanes and marked off parking lanes on either side so the turn is fairly easy.

What wasn't easy was telling the car coming south toward him on 'B' Street that the truck would take up the whole intersection and that the ends of the poles would swing wide as he turned. Earl signalled and waved, but neither the elderly driver nor his wife in the front seat beside him understood. They just kept on coming, never even slowing down. They drove right under the ends of the poles.

As the back right tires of Earl's truck bumped down off the curbing, the load bounced and the shifting logs plopped a huge load of mud right on the center of car's roof.

If the old couple ever knew how close they came to disaster or what had gotten them on top, they never said anything, at least not that Earl could hear. They never did slow down.

Up in the woods, you don't expect to see much other traffic. Earl was driving up in the timber above Clear Creek. He had a whole load on and was hauling pretty fast up there on the logging road alone, he thought.

But an elderly couple had come to visit someone on Clear Creek and gotten lost. The old gentleman was so terrified when he saw Earl's truck barreling down at him, that he jammed on the brakes, huddled over the wheel and closed his eyes. If he'd gone to one side or the other, Earl could have gotten around him, but he stopped right in the middle of the road.

Earl braked hard and held his jaw tight. The truck stopped inches from the old man's front bumper. Earl jumped out of his truck and ran up to peer in at the two terrified people. They were unhurt, but frightened out of their wits. Earl finally talked the old man into backing to one side of the road so he could maneuver the truck around them, then watched in his mirrors as they followed him out of the woods down to the road.

Bud Betts (also known, to his dismay, as "Betty Butts") once was riding out of logging camp with Jake Taholsky driving "hell bent for election." They came screaming around a curve and met head-on with a log truck. A loaded log truck has enough momentum that it can't usually stop on a quarter, let alone a dime, so Jake knew if they hit, he and "Betty Butts" would be smeared on trees for a hundred feet all around.

With nowhere else to go, Jake veered off into the brush, driving between the trees and praying there weren't any stumps. There weren't. That time. Jake cut a half circle swath in the brush around that truck, came back onto the road -- laughing -- and drove fast all the way home.

Bill Franck was another one who laughed when things happened to him. He needed brakes for his old pickup truck, so he went to Alfred, who was keeping books for the Nordgren Timber Company at that time. Alfred paid him, and Bill went into town for the brakes.

The next day, Bill came back and told Alfred he hadn't gotten enough. So Alfred sat down and went over the ledger with him, showing Bill when he'd worked and how much he'd made day by day. When they finished, Bill looked at him and laughed.

"By God, I owe you money!"

As hard as it is to keep a logging truck maintained in running order, drivers know there is more profit in carrying a heavier load each time, rather than making extra runs with lighter loads. But the enormous weight of the loaded trucks causes wear and tear on the rural roads.

Government inspectors are hard put to keep the log loads within regulated limits. They often go out with jump scales onto the back country roads to catch loggers coming in to the saw mills. One of the Lydas -- a younger relative of high-climber Ernie Lyda -- was a brash young guy who always had a terribly heavy load. They caught him on one of those back roads and put him on the jump scales. His load of logs was so heavy, they screamed at him, "Get off'a there! You're breaking our scales!"

"W" used to drive a log truck. Once at a Weigh Station the officers told him he was 10,000 pounds overweight, and he nodded and said, "That's about right for me." But instead of waiting for them to write up his infraction (there was a stiff fine for that much overage), he snatched up his paper and took off in his truck. They never did catch him.

("W" did other crazy things when he drove. Sometimes he powered his loaded log truck through highway construction sites at 65 mph. He loved to see the sign men and workers scatter. He used to brag that sometimes it was ten minutes after he'd passed through before any of them recovered enough to go back to work.)

Earl was driving a pretty heavy load of logs on Stringtown Road west of Forest Grove. He looked east out across the valley floor and saw exactly what he was hoping he would not see: a government scaler driving northwest on Gales Creek Road. As soon as Earl saw him, he knew that the scaler had seen him, too.

The scaler was hurrying ahead to Watts, where he could turn onto Stringtown Road. That would put him just behind Earl. With the load Earl was carrying, he knew it wouldn't be long until the scaler caught up with him.

But unknown to either one of them, another logger had just come down off David Hill Road and turned onto Stringtown between them. That second logger also had a pretty heavy load of logs, and the scaler pulled him over, thinking it was Earl.

The unlucky logger had some overage, too, so he got a ticket. But the extra weight was not nearly as much as the scaler had estimated when he saw Earl's truck across the valley. Realizing he'd caught a different truck, the scaler then took off down Stringtown Road after Earl.

"But by the time he caught up with me, I'd just got to the saw mill and enough of my load was already in the mill pond that he couldn't ticket me for the overage," Earl grins. "That was one time."

Once a trucker got his load to the sawmill, he'd lay a "brow log" there at the edge of the hill above the mill pond so his truck wouldn't sink in the mud as he was dumping his logs.

The logs were rolled off the trucks and down the bank into the water where mill workers manipulate the floating logs, sorting them by length and type needed next to be put under the saw mill blades.

Trucks weren't the only way logs were hauled out of the woods. Sometimes the timber company would build and maintain a whole railroad system within their enormous forest tracts. Shays were special railroad engines that could work up a hillside into mountains. The wheels were lower on one side than on the other.

The timber company would build spur lines up to the different cutting sites to connect with their main line that went to the saw mill, beside the pond.

Floating logs are so much easier to handle and move around, timbermen try to use any available rivers or lakes.

Old man Benson logged north and east of Hillsboro on the Chapman Burn and sometimes out of Scappoose, up along the Columbia River. Benson was one of the first to build a "cigar raft" of his logs. He floated those rafts all the way to sea and then down the coast towed by a tug. Sometimes there were a million board feet of timber making up a raft.

In calm waters all that is needed to make a raft of floating logs, is to chain together, end to end, the logs at the perimeter. As a raft, logs could be transported much more cheaply down a river to the saw mill rather than loading and hauling them truckload by truckload.

Usually, though, there is no suitable river nearby. Sometimes the logging site is so remote and so mountainous, that the only way to get the logs out to the mill is by air-lifting them with helicopters. But that is a very expensive way to haul logs. The quality of the wood has to be very fine indeed to make air-lifting them to the mill a paying proposition.

At the saw mill, a scaler measures the length and diameter of each log to figure out the board footage and how much the logger is to be paid for the timber he's brought in.

Les Friese was one of the very few scalers who always worked in a shirt and tie. He'd go right out onto those logs in the pond with his scaling stick and start at one end and walk along those logs slick as anything. He'd be measuring as he walked, so when he came out the other end he could tell just how many board feet there were. And he'd be right, too.

For a long time Forest Grove had only one saw mill, and most of the logs were brought in by train. Each train had about eighty cars, each approximately forty feet long. The men loading and unloading all that timber were called "log punks."

Piles and piles of stacked logs, called cold decks, stood all around the mill. Most of the mills kept sprinklers watering down the logs all through the summer so they wouldn't dry out and crack, or catch on fire.

Workers in the saw mill are called "sawdust savages." Working around the saws is dangerous, but even a man working at unloading the sawed pieces (working the "green chain") could lose an arm or an eye.

The green chain is a set of two linked parallel, moving steel chains about four feet apart. The newly cut pieces of board hang suspended between them. The green chains move continuously, and the workers have to grab the moving pieces off and sort and stack them according to size.

"Saw milling, that was a rough job," a logger said. "You worked so hard but didn't always get paid for your product. They used to joke that the only way to make it in a saw mill was to marry a rich wife."

One old timer knew everything about logging, from cruising to green chaining. He could be heard day after day complaining about how badly things were run.

"If I was in charge, I'd fire them all!" he growled. There was a beat while his face changed as he thought about the consequences of doing just that. Then he grinned. "We'd be in a hell of a mess then, wouldn't we?"

It's said that Gar, who owned a saw mill, used to be a cowpuncher and a bronco-buster in Montana. He would tell the classic story that makes loggers grin:

A little logger walked into a bar and, after a few drinks, hollered, "I can whip anybody in this here bar." When there was no response, he drank a little more, then yelled, "I can whip anybody in the whole town." Still no one reacted, so the little logger drank some more and boasted, "I can whip anybody in the whole Northwest!"

At that a huge, raw-boned logger stepped over and punched him so hard he flew across the room and smashed into the wall on the far side. As the little logger staggered up and rubbed his jaw and checked for loose teeth, he grinned.

"Guess I took in too much terri-tore."

Chapter 8

Every Failure Has Something In It

The Nordgrens logged for years using for power their own straining muscles and that of their faithful horses (even when they weren't diapered). But gradually mechanization eased some of their labor. Men were freed to produce more lumber, just as families were freed by the automobile to go longer distances in less time.

Alfred loved his new Nash. He did not go far or often, but when he went, he went fast. Eighty miles an hour, sometimes, at a steady pace. People asked him why he wasn't afraid of having a tire blow out.

"No tire is going to blow out," he scoffed. "It's going around so fast it doesn't have time."

He drove pretty fast up in the woods as well. Not eighty, but fast enough to raise a lot of dust.

Alfred once stopped to pick up a hitch-hiker. The man ran up to the car and opened the door. But when he looked in and saw all the dust, he backed away and closed the door, murmuring, "Ah, no, thanks, I think I'll wait for the next offer."

The family wished one of their young nephews had waited for the next ride instead of going drinking and driving with a high school buddy. He rolled down the pick-up's passenger side window and leaned way out, hollering and laughing as the driver swerved at signs -- until they side-swiped a stop sign that nearly decapitated him.

"Who's to say who's to blame?"

Loggers sometimes got so drunk on their weekends off that they'd miss the last train from Portland out to Forest Grove, a twenty-five mile trip. Earl remembers when the road over the West Hills into Portland was made of board planks. There wasn't much traffic, and little chance of a logger hitching a ride out to the woods. But the hapless logger knew he had to get to work on time early Monday morning, so there was nothing to do but hike it.

Charlie F would get drunk nearly every weekend. Monday morning he'd report for work so high-strung and nervous, he'd fall a tree and run a hundred feet to get out of the way. Later in the week he'd know which way it was going to fall and just step a couple feet out of its path.

E.B. carried a cooler in the back of his log truck. You could tell which road he went up by the empty beer cans along the side of the road.

An old man who was stone deaf drove all around town with one foot on the gas petal and the other foot on the brake. You could hear that poor car grinding and revving for blocks around.

The volunteer fire department in Dilley had an emergency. The men were clinging to the fire engine roaring down old 47 Highway, their hats turned around backward and the wind in their faces. Then they were passed by an ancient, wheezing commuter bus. The volunteer firemen dropped their eyes in humiliation.

Some Oregonians claim Squatters' Rights to the fast lane, but it wasn't always that way. Many of the old timers learned to drive in their fields, so they naturally took the middle of the road.

A Swiss man and his wife and daughter owned a dairy farm near St. Helens, a little Oregon town along the Columbia River north and west of Portland.

Two young Swiss brothers came to work on the farm for them. Those boys were real musicians. They played clarinet and accordion and were really in demand to play for Swiss get-togethers in Helvetia and in Cedarville Park (near Gresham on the way up to Mt. Hood).

On the way home from one Swiss party, the family was involved in an automobile accident that killed the dairy farmer.

The older of the Swiss brothers married the farmer's daughter and the younger brother married her mother. An unusual situation, perhaps, but it worked well.

Earl bought five hundred acres of timber from a college teacher, an astronomer who had more land than he could manage. His work at the college was time-consuming and mentally demanding. As a change of pace, he liked to work in his fields on weekends. He was out plowing one afternoon when his tractor tipped over on him and killed him.

Tom Laman had a Model T Ford, but he didn't know much about driving it. He came putting along with Nat Hoover beside him. But Earl and the other kids had found an old hornet's nest and set it right in the middle of the road.

Poor Tom swerved his Model T up onto a bank to avoid hitting the nest until he was at such an angle that the car tipped over.

"We kids helped him tip the Model T back up on its wheels. It was so funny. Tom couldn't really be mad, but he had a pipe in his mouth, and he was shaking so bad that pipe was bobbing up and down."

Not all childhood pranks are written off as funny, however.

"When I was young I was always threatened with reform school if I ever misbehaved," Ivan Marble said. "Since then, I've always hated the idea of 'reform.'"

"That must have been the punishment of the time," Lori Andresen agreed, "because I was threatened with reform school, too. I hadn't any idea what it was, but I knew it was bad."

"I was never threatened with reform school," their friend Walt said, "but I knew I'd really been a stinker when my mother said she'd trade me for a yellow dog -- then shoot the dog."

Don Wilhite never could explain how his mother (whose only complaint was, "I've just got a hitch in my get-along") managed to bring up her family on their farm. She could sure keep discipline, though.

"To this day I can't understand how my mother could come up with a vine maple switch in the middle of a forty acre strawberry field."

The Tohalsky boy from Russia belonged to a religious group with the Goffs. Every Saturday night they'd gather on the street corner across from the Forest Grove Bank. About twenty of them would stand around listening as old man Goff preached. Sometimes the Tohalsky boy did the preaching, so well he was converting the Vaandering boy. But old man Vaandering wasn't pleased.

"That Tohalsky boy," Vaandering grumbled, "he make me boy crazy."

The elder Mr. Goff wanted to build a new church so he went to an architect and asked how much it would cost to draw up the plans. When he heard the price, he scoffed, "Oh, hell, we could build the whole church for that." So they did.

If religion wasn't always the whole answer, perhaps neither was education. One of the high school teachers was an excellent musician and golfer. Everybody liked him. He was friendly and a good talker. He also had an enormous appetite. At a restaurant in town he'd order a huge steak with all the trimmings. He'd eat all of it and then order another one, and eat all of that, too.

He was such a nice guy even his wife didn't know what he was up to. No one caught up with him until he rented a car and didn't take it back. Auto leasing companies tend to go after people who don't bring their cars back.

During the investigation it was discovered that he'd been taking high school kids to Eugene and other Oregon towns to cash unauthorized checks for all sorts of companies. It took them a while to find out just how deep he'd gotten himself involved.

Just off Gales Creek Road near where David Hill Road comes in, there used to be a popular picnic spot called Rippling

Waters Park. It was Stew Petit's favorite place to go diving. When Earl was eight or ten years old, he was picnicking there with his family. A lot of people were swimming and some were doing fancy dives off the creek bank into the swimming hole.

There was a man there who couldn't swim, but he got so excited seeing all the others diving that he decided to try it, too. He dived in, but he never came up. They dragged him out and draped him over a barrel to pump the water out of him, but he never woke up or breathed again.

E.B. had been drinking with buddies one night when somebody came up with the brilliant idea of stealing a tom off old man Bates's turkey farm. E.B. chuckled and rubbed his hands, ready for some fun.

One of the guys drove them in his blue sedan out to the road west of Forest Grove. At the edge of the turkey farm, E.B. got out and leaned back, saying, "You stay here and keep the motor running. I'll just run in and catch one of those birds."

E.B. climbed the fence and started into the fowl yard. Those turkeys raised a terrible rumpus. So much so, that Bates heard them and ran out and piled into his own blue car and drove around the road to the field. E.B.'s get-away buddies saw Bates coming and drove away, but E.B. was so busy tackling a big tom that he didn't see them go. Bates just pulled his own blue sedan into their parking spot along the side of the road and waited. Pretty soon, here comes E.B. dragging that huge tom turkey still twisting and flapping its wings.

"Come on, let's get out of here!" he yelled, getting into the blue car. "With all the ruckus, old man Bates is sure to have heard us by n... Oops," he stammered realizing finally that it was the wrong blue car and the wrong driver.

"Ah, Mr. Bates, this is going to be kind of hard to explain..."

Bev Olson and a dear friend got so excited giggling and talking together in the kitchen that they lost track of time and the pressure cooker blew its cover. There was spaghetti sauce all over the ceiling, which would have been hard to explain if everybody hadn't known what those two were like when they got together.

Joe Loomis was a wonderful guy. He sold heating oil in the Forest Grove area for years. Sometimes people would run out of oil in the middle of a cold night and call him. Joe would get out of bed and take them what they needed.

But Joe had a bad heart. He had a massive heart attack and was in the hospital while Earl was hospitalized for months after an automobile accident that crushed his hip. Earl heard Joe had been brought in, so he wrote him a note: "No wonder you had a heart attack. Look at all you've done for so many people over all those years."

The next day the whole Loomis family crowded into Earl's room to thank him for touching Joe's spirit and giving him the courage to fight on.

Courage is an everyday commodity in most small towns. One old farmer up on Thatcher Road lost his leg below the knee. When he got out of the hospital, he took the yellow plastic delivery tube for the Oregonian Newspaper and pulled its open end over his stump. He tramped around on that thing just fine.

The Snyder brothers, Walt and John, went to the hospital to visit their father. They'd never been in a hospital before. They'd never been in such a big building before. They didn't know to ask at the front desk where their father was.

"We must'a looked in 300 rooms, but we never did find him."

Dr. Bump's aged mother was in the hospital when Earl was there, and for almost as long. She was getting pretty disheartened. One day Earl ordered a dozen roses sent to her with a card asking her to go to the dance with him Saturday night. Oh, she got such a kick out of that.

Dr. Bump has an effective bedside manner sometimes. He once told a new widow, "You're not broken up because you lost your husband. You're upset that your lifestyle and income will be going down." It did shock her into looking at things a little more clearly.

He did have pity on one patient, telling her she was "...almost as bad off as a woman with eleven kids and a drunk for a husband."

Rich H was in the old hospital in Forest Grove, on the second floor. All his friends came to see him and snuck in bottles of beer or liquor. They didn't want the nurse to see it, so they hid the bottles outside his window in a vine clinging to the wall. It got so heavy, the whole vine tore away from the wall and everything fell down with terrible crashing and smashing of glass. They all thought that was so funny.

Children were warned to stay home on Saturday nights. Loggers had worked hard all week and on Saturday nights they played hard, they drank hard, and they fought. Harrison Heisler got his nose severely bitten. "I got in a fight with a guy who was more of a dog than a man."

To keep a semblance of peace, Forest Grove had a big Irish cop named Mickey, who carried a sack of shot to hit the men on the back of the head with to get their attention. Mickey was walking down near the Star Theater one Saturday night when

Cannibal Beard came staggering around the corner from Council Street. He was drunk and loud and obnoxious, and paying no attention to Officer Mick.

"Cannibal! So help me, Cannibal, I'll make you listen to me," Mickey warned and lashed out with the terrible sack of shot. He hit Cannibal full on the back of the skull, but Cannibal stayed standing. The shot bag didn't fare as well, though. It broke, and the shot scattered and rolled all over the street.

Cannibal turned and shook his head slowly, then reached for Mickey and picked him up bodily and threw him through the plate glass window of the department store across the street.

Cannibal Beard was huge and tough. Norm Smith was far smaller, but the two of them were great buddies in the late 1930's and 40's. Somebody asked Norm if he was going to war with Cannibal Beard.

"Yep! He's gonna kill 'em and I'm gonna count 'em."

Smith was such a comical guy. He could make you feel sorry for the mice, "they come out of the wall with such a hungry look."

Old Charlie F had all his teeth pulled out, but he managed to get to a party being held at a log cabin up in the woods. Charlie played the mouth organ, so drunk he didn't even miss his absent teeth. He had the best time. He went home that night and flopped down on the bed, breaking it down on one side, but he and his wife slept on it all night tilted like that and never knew the difference.

Sometimes Charlie would finish his work at night and come sit down at the kitchen table and say to his wife, "Let's mosey down and slop up a few." And they did.

At the McEvoy Dance Hall they didn't sell liquor, so you had to bring your own bottle if you wanted some. And most people wanted more than some. But the night Earl went with friends, nobody had realized that was where they were going to end up, so nobody brought along even one bottle.

They sat down at their table and looked around at the others drinking and having a good time. In fact, at the table next to theirs, one man had a full bottle. They watched him hide it in his coat while he went out on the dance floor.

"Did you see that?" Willy asked. "Keep an eye on him, will ya?" While the man was dancing, Willy sneaked over and stole the bottle from the man's coat. He opened it and served drinks all around.

When the guy came back, he quickly discovered his bottle was missing. He searched all around, but no luck.

"What's the matter?" Willy asked him.

"Somebody stole my bottle."

"Oh, that's too bad," Willy commiserated. "Say, why don't you just pull up a chair and have some of ours? We've still got almost a full bottle here."

So the guy pulled up a chair and joined them, drinking from his own bottle and thinking these were pretty swell fellahs.

Waldo R went with friends on a fishing trip. When Waldo arrived with a whole load of beer and a single loaf of bread, he was greeted by cat-calls: "What in sam hill are you doing with all that bread?"

E.B. and Pete L went up to Alaska and logged for Pat Soderberg. Soderberg wouldn't allow any liquor at all in his logging camp. E.B. and Pete had gone to town and were coming back in their little boat, sneaking a case of whiskey back with them. Still a hundred feet out, they saw Soderberg standing with his feet apart on the beach, watching them.

"Uh - oh." They tossed the case of whiskey overboard right where they were and came on into camp. There wasn't much Soderberg could say.

But later that night when the rest of the camp was quiet, both E.B. and Pete were seen out in the little boat again about a hundred feet off shore hunting something by lantern light.

Before they went up to Alaska, E.B. and Pete logged in the hills west of Forest Grove. One time they came to town all liquored up and walked into the tavern with their calk shoes still on. They started to play shuffle board, but they weren't very co-ordinated, so they didn't do very well.

"It's that dang smooth surface," they groused. But that was easy to fix. They went skating up and down the shuffle board alley gouging the polished surface with their spikes.

Another time E.B. and Pete went to a tavern directly from deer hunting. They were already pretty well lighted up in the woods so E.B. carried his huge bow and arrows with him into the tavern. As he entered, the stuffed deer head mounted on the wall caught his eye. He whipped out an arrow and shot that deer head right between the eyes -- again.

Jim S had three martinis in a row at the Jungle Room. Then he started thinking that his wife would be mad if he didn't call and tell her he was going to be late. He got to the phone booth all right, but he slid down with the receiver in his hand and fell asleep on the floor of the booth. He never did get finished dialing.

"Every failure has something in it," Earl says, "and that's called discouragement."

Earl's theory is that alcoholics are irritable when they're sober because they feel lousy. Convivial drinking is a way of sharing -- or denying -- the guilt. They can get pretty offended and angry if somebody refuses to drink with them. But sometimes it's hard to keep up, even if you wanted to.

There was a club off Barbur Boulevard that had a swimming pool with a glass wall as part of the decor of the restaurant and bar. They had beautiful gals in attractive swimsuits diving in and swimming around under water for all the patrons to watch. One night one of the male patrons went into the Men's room and took off his clothes and dived in with them.

Earl sings his own version of "Lara's Theme":

> "...out of the green and gold
> and there'll be girls,
> more than your lap can hold."

Earl's brother Fred had charisma. Anywhere he went, people gathered around to listen to him talk. Fred enjoyed life so much, he could laugh and make those around him enjoy themselves, too. He was big, handsome, irrepressible. One New Year's Eve, Fred kissed every one of the ladies.

"It felt so good, I did it again!"

For years he seemed to have a charmed life. He had tickets for a flight, but was bumped at the last minute when the plane was over-booked. The plane crashed.

John Hagen had a beautifully kept farm that he sold for $10,500 and thought he would be on Easy Street for the rest of his life. He and his wife moved into Forest Grove where they built a brick tavern next to where the McDonald's is now.

He sold the best beer, which he kept in big kegs. Every night he cleaned out the pipes with hot water, so his beer always tasted so fresh people loved to come to his tavern to drink.

Hagen's wife Lena made excellent sandwiches, and people loved to eat there, too. Normally John was an easy-going guy, but sometimes if he drank he got depressed and irritable.

Earl stopped in. "Boy, it's a nice day," Earl said, but Hagen scowled.

"It's gonna rain pretty soon."

Earl closed his mouth and watched John return to the argument he'd been having with his wife. Hagen was usually a quiet man who kept things to himself. But Earl hadn't realized how much he kept to himself or for how long until he heard Hagen accuse Lena of having a roving eye.

"A roving eye? Me? When did I ever look at anybody else?"

"That guy in Oslo!" Hagen stormed.

"Oslo? We haven't been in Oslo for forty years," she laughed.

Old age has its own past -- and its own present. 98 year old Bud Pietsch and his

92 year old wife bake their own bread from wheat they grind themselves. Bud is considered too old to drive his car, so he rides his bicycle -- and carries the groceries home in the bike basket.

Mrs. L lived north of Forest Grove. She had the country woman's dislike of strangers and had taught the family goat to butt people. Most of the time she thought it was pretty funny. She was out in the field picking blackberries when two men came to the farm to try to buy junk for scrap metal.

Mrs. L heard them coming and raised up from where she'd been picking berries. The goat came dashing across the field and butted her flat into prickly berry bushes.

The visitors, instead of helping her, roared with laughter. Mrs. L picked herself up, scooped up her buckets and stormed into her house. From her window she glared out, then laughed, too.

The men were so busy guffawing, they hadn't realized the goat was making another pass. It flattened them, too.

An older auctioneer with false teeth was doing a charity pie auction in Forest Grove, holding the microphone in one hand and each pie for sale in his other hand. He got clucking away so loud and so fast into the microphone that his false teeth flew out of his mouth, falling forward and down. Both hands were full. He had to jolt his face forward fast to catch those teeth in his mouth again before they hit the pie.

A neighbor who owned the shore lot next door to the Nordgrens' south of Cannon Beach had been a prize fighter in his youth. Determined not to let his body grow old and flabby, Jarvis exercised regularly, including rowing. Wearing a wetsuit because of the cold of the Pacific along the Oregon coast, he would go out on the ocean in his little boat.

One particularly cold and miserable day, Mrs. J tried to discourage him from rowing around the light house off Cannon Beach. She was worried. The waves were high and the weather threatened to get worse, but J was determined to go out anyway. Mrs. J asked Earl to go with her to watch from the beach.

J made it around the light house rock and was nearly back to shore when a huge wave came in that lifted J and his boat to its crest, then tossed them aside. J floundered in the cold water, trying to stay afloat, but another huge wave came up right behind the one that had knocked him out of the boat.

As Earl and Mrs. J watched, he was sucked down, then lifted higher and higher until he was finally dropped -- back into his boat!

Chapter 9

Taking Care of Business

Charlie Zumwalt was working for a family whose little girl just loved him. Imagine his -- and the family's -- surprise when she announced one day, "Let's sell Charlie and buy a bull."

Charlie worked in a service station when he first married Alfhild Nordgren. He worked the second shift. One day the guy who worked first shift left the station early and locked up, but he put a note on the door:

"The key is under the mat."

Charlie got a big kick out of that, especially when he looked and the key was still there.

Not everybody defines honesty that same way, however. One month the man who was leasing a house in Forest Grove from Earl drove up in a brand new car to say he couldn't pay the rent. The next month, he skipped out. When Earl garnisheed his wages, he thought that was unfair.

Earl's face went drawn and tight as we drove mile after mile along the fenceline of a Colorado ranch.

"All the work and all the troubles in the world are right there on that ranch," he said. "Land of your own and it's up to you to make it work. Make it pay. What I hated most was the awful guilt feeling when I left something unfixed."

Colorado has its own pace of business.

Earl was glad when he saw the large sign painted on the upper story of the exterior brick side wall of one of the buildings on the main street of Steamboat Springs, Colorado. "EXPERT Shoe Repair here," it promised, and Earl needed the heel fixed on his shoe. But no one inside the store knew what he was talking about. "No one here does shoe repair." They called up the store owner.

"Oh, that sign? It was put up there twenty, thirty years ago. No one's bothered to take it down."

Colorado and Wyoming also have their own brand of entrepreneurs, especially in the pioneer days before Jim Hollon brought his bride, Jane Nordgren, home to Laramie.

Far out in the back country, Axel Palmer harnessed four wild horses together and hitched them to a cart. By the time Palmer had driven them all the way into town, the four strong individual horses pulling against each other finally wore them all out and broke them to harness.

One time a neighbor asked him for a ride to town. Axel set his jaw, thinking. At last he said he'd take him, but warned there'd be no stopping along the way. The neighbor agreed.

"I got no need to stop 'tween here and town," he drawled.

On the way in, though, the team was particularly wild. They were pulling Axel this way and that. Finally the cart hit a bump and the neighbor went flying off, ending up sprawled on the ground. Axel turned and looked back, but there was no way he could rein in the horses to go back for him. The neighbor was on his own.

Especially in the Depression, so many Northwesterners were on their own. There were a lot of haywire outfits that scarcely made things hang together. They were called gyppo loggers, and their equipment was often barely usable, and their methods sometimes barely honest.

Many of the other loggers had little respect for them and would speak of the gyppos as "a little bit lower than the bottom of the barrel."

"Gyppos can crawl under the belly of a snake -- with a top hat on."

There was even a poem of sorts about them: "Haywire outfit on a hemlock show.

Skin back easy and go ahead slow."

Once when Earl stopped at a small airport in Oregon to fuel his helicopter, he watched a brash young couple fly in with a beat-up Piper Cub with different paint for different sections of the fuselage and wire to hold most of it together.

"Now that's a gyppo airplane," he chuckled to himself, but he gave them a hand pushing it up to fuel pump. "You two out for a joy ride?" he asked, wondering why anyone would risk his life in such a battered aircraft.

"Oh, no, we're going to Alaska," they told him. Earl shook his head as he watched them take off, barely able to lift the plane off the runway.

The old Dodge truck Truey Snyder used to haul logs was so worn that Truey (short for Trueman. His brothers were named Worthy and Sincere) had to carry with him a bag of piston connector rods so he could pull over and replace one when it broke.

But Truey worked hard, gradually able to buy better and better trucks, even though a medical problem with his back was slowly bending him like a comma. He kept at it and kept at it, until at last he could afford the best -- a beautiful, big Kenworth.

And just when he'd achieved his goal, economic times got hard. Competition got more and more desperate. People were willing to haul logs for next to nothing, simply to have a job. It was hard to make anything at all. He'd spent so much time and effort buying that Kenworth, thinking it would bring prosperity, but then he was barely able to hang on to it.

Alfred had taught his sons to deal straight-forward, to buy and to sell on a cash and carry basis, or play the sucker every time. Earl had done a job for Walt Hayden and Walt hadn't yet paid him. Walt was working down by the river, so Earl walked down there to get what he was owed.

But Walt didn't have the money. He explained carefully just how he planned to juggle this, that, and something else and pay up when it all came together later on.

Earl was getting hotter and hotter under the collar and began arguing with him as he could hear in his own mind his father's warnings never to get taken in by promises like that in business. Walt could see he wasn't satisfying Earl, so he began again to explain the plans he had, but Earl clearly wasn't buying it.

Finally, exasperated, Walt drew himself erect there by the river, put his hands on his hips and asked, "Well, how would you do it, if you was me?"

Earl hemmed and hawed, but there wasn't anything else to be done but wait. Walt Hayden did eventually come through as promised. He was one who always did.

Walt Hayden had a knack. He was out driving when he saw a large flock of turkeys. Now he happened to know of a guy who wanted turkeys. So Walt drove up the lane to talk with the farmer. They agreed on a fair selling price for the entire flock. But Walt knew the guy looking would pay a couple thousand more than that.

Walt was in the bank in Forest Grove one day not long before Christmas. The bank president caught up with him in the lobby, wanting Walt to give a sizable contribution to some charity. He kept pushing him. Right there in public he needled and cajoled, actually setting out a check for Walt to sign.

"How much money do I have here in this bank?" Walt asked, knowing it was close to a million dollars, and knowing that that represented a sizable chunk of the bank's assets. It was an enormous amount in those days. "Well, you just make out that check for everything I've got in this bank."

Mouth open, the banker watched Walt carry that check out to deposit it in the bank across the street.

When Walter Ball came to work for the Nordgrens in the woods, Alfred asked him what he did. "By trade," Walter said with a straight face, "I'm a bank robber, but it looks like you're full-handed."

"What good are friends?" loggers groan. "You can't borrow money from them."

One banker is supposed to have looked up when a hauler came in, hat in hand, wanting a loan to buy a new truck.

"Lost control of my log truck," he drawled. "Smashed it up pretty good."

"Didn't you have insurance on the truck?" the banker asked, and the hauler nodded, yes. "Well, then you just go on over to your insurance agent and have that old truck declared a total loss. Then they'll get you a brand new truck."

The hauler's face brightened. "Say, I wonder if'n that would work on my wife?"

A logger who bought a lottery ticket was asked what he'd do with a million dollars tax-free. He thought a minute, then muttered, "I guess I'd just keep on logging until it was gone."

One man who worked in the woods wasn't very good at logging. He poked and fouled up until finally his exasperated boss told him to take a hike back to town right now as his way of telling him he was fired. He wasn't even to wait until the crummy was ready to take the men in at the end of the day. "Go count all the railroad ties on the way to town," he told him.

Well, the man did just that: he counted every last railroad tie between the camp and town, then put in for wages for the time he'd spent counting them. The company refused to pay -- until the courts told them they had to. The man had done what his boss had told him to do and was entitled to be paid for his efforts.

Earl first bought up his father's land, then gradually over the years added nearby parcels and farms as they came on the market. Sometimes he would log part of a farmer's land for him, then buy it when the farmer retired. In each transaction he tried to deal with the integrity his father would expect of him, and often the man Earl dealt with on one occasion would remember and deal with him again years later.

One old man had Earl log his land for him, then balked when Earl brought him the check as they had agreed on. "I've been thinking it over and I think I should have gotten more," he complained.

"How much more?" Earl asked.

"The prices for logs went up," the man said, suggesting a higher figure.

"They did," Earl agreed and they settled right then and there on a fair share of the extra profits. Years later, Earl bought his farm. He wouldn't sell it to anyone else.

A couple had some land near Clear Creek, outside Vernonia, that they wanted to sell. Earl went up and looked over the parcel and came back to ask them what they wanted for it.

"$6000 and nothing less," they said. "We've checked the banks and people you've done business with and we know you drive a hard bargain, but we want $6000."

Earl knew it was worth every bit of that, but he knew they wanted to feel they'd driven a pretty hard bargain themselves, so he tipped back his cap and squinted his eyes as though he was considering it.

"You sure you wouldn't take just a little less?" he asked, looking pained.

"Nope. $6000."

Finally Earl agreed to give them what they wanted. Later he took more than double that amount's worth of timber off the place and still had the land to re-sell. But he never told the couple.

Earl bought some land from W. When all the dealings had gone through, he took the check up to W's, but he wasn't home. So Earl gave the check to Mrs. W. A week or so later, W saw Earl in town and asked whether his check had come through yet.

"Yeah, a week ago. I took it up to your place and gave it to your wife," Earl answered, surprised.

"She'll never give it to me," W said, shaking his head. "You'll have to get it back from her or write me another one."

Earl couldn't quite believe him, but he did go up to their place and talk with the wife. It wasn't until he explained several times that the check was made out to Mr. W and she could never cash it that Mrs. W starting even looking doubtful that she'd keep it.

"Besides, unless you give it to me now," Earl added, "I'll have to go to the bank and have them void that one so nobody can ever cash it. Then I'll make out another one and give it to him direct."

"You'd do that?"

"'Fraid so," Earl said, and the woman relented finally and went and got the check and gave it to Earl. She still wouldn't give it to her husband.

Another time when Earl was cutting timber on their land, W and his grown-up son came up to where he was working.

"Prices has gone up for timber," W growled.

"That's true," Earl agreed. "By our agreement, I'm cutting on your land and paying by stumpage," he explained again, indicating to both father and son how much he'd agreed to pay per board foot, and how much he'd already cut plus how much more he expected to.

"Yeah," W insisted, poking the toe of his heavy shoe against one of the stumps, "but prices went up. I could have got $100,000 for this."

Earl looked between father and son, then tipped his cap back on his head.

"I'll give you $100,000 for it right now, if that's what you want," he grinned.

But the son had been working over the figures in his own mind as his father growled. "Pa, Earl's already agreed to pay you more than $100,000. You better shut up and leave it alone."

Alfred Nordgren once bought a right-of-way through a man's property. The family was really poor. The man bought his wife a kitchen stove with the money, growling as he gave it to her, "Here, now just shut up for the rest of the summer."

In business, as in life, things twist around. One time Earl had to get a right-of-way up through B's property to timber he had bought behind it. There was a log road in already. It was even already rocked. Earl offered B $4000, but B realized Earl couldn't get to that timber without the right-of-way, so he demanded $10,000. He wouldn't compromise.

Earl's attorney advised him to go to court and sue for the right-of-way. The law had recently been changed because of right-of-way robbery where some big companies were squeezing out small timber owners by not letting them get back to their land.

In court, Earl was awarded the right-of-way. He could use road without paying anything to B for it, not even the $4000 Earl had offered in the first place. Not only that, but B was not permitted to use the road. And B had to pay court costs!

Earl aches for the loggers out of work in their struggles with those dedicated to preserving the wilderness. Although they, too, want to keep the beautiful Northwest as heritage for their families, many loggers feel that some environmentalists have demands that are way beyond good sense and that they refuse to listen to reason.

"Don't confuse me with facts. I've already made up my mind."

Their own economic hardships have made some of the loggers bitter. One complained about trying to deal with an unreasonable environmentalist and was told, "It's no use trying to trade smells with a skunk."

After ten years of his own retirement, Earl realized, "Loggers don't want to do anything else. They don't want to be retrained. They work so hard all their lives without having time or energy left over to do other things. So when they retire they have no other interests. They sit on the porch and snap at their families -- and wait to die."

"I worked for seven years straight in the ship yards without a day off," Earl said, "because I knew it would take that kind of sweat to get ahead and have enough money to be able to do what I wanted to."

Over the years Earl bought up property until he finally owned over five thousand acres of prime timber. When he decided that he'd been at it long enough, he made a deal with Willamette Industries. The firm president wouldn't let Earl take stock in Willamette Industries as payment, perhaps not wanting to deal with the power that would go to someone with that chunk of stock. He would only pay in cash. But soon he, too, retired and moved to a country club in Rancho Mirage, California. He died very shortly thereafter.

"I sold all that land and never went up there again," Earl said. "I flew over it a couple times in the helicopter, but I never looked back after over forty years tramping over every inch of it."

"Never in my whole life have I felt I'd done anything big. Not even when I held that check with all those zeros in my hands." But then he grinned. "I put that check in the bank until I could get it all distributed to owners of the Nordgren Timber Company.

Precious middle daughter Mary Jo was back in Forest Grove after time in San Diego while her high school sweetheart husband Rich Clyde served in the Pacific on an aircraft carrier. They used their part of the money to build a home in a corner of the land between Alfred's and Earl's that became known as the "Swedish mafia corner."

"When I took Jane's to her in Wyoming," Earl says, "I went with her to the bank in Laramie. The president came out and told us they wouldn't be able to do business.

"'Insufficient funds,'" he told us, then hastened to add, 'on our part, not yours. We can't handle a check of that size.'"

Earl drove with Jane to one of the larger banks in Cheyenne.

"We could see people scurrying around to check with the manager. Pretty soon we saw him tilt his chair and look out to peek at this guy who'd come out of the woods."

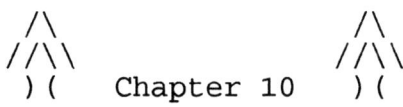

Chapter 10

Legends of Axel

Axel Erickson was a wild-eyed Swede, almost 6 feet tall, two hundred pounds of muscle, with thick wrists, Pop Eye biceps, a broad face and enormous hands. He walked fast, worked fast, told the world what he thought, and the world, in turn, has a few stories to tell about him!

Axel had his own brand of X-ray vision that could see right through bureaucratic nonsense. He felt most of the regulations of the logging industry were a power play, and he wasn't about to let anybody tower over Axel Erickson.

He could bluff. And he was honest, and dishonest, enough to get away with it. He's said to have had about $10 million when he left Oregon and went to California. But it was in California where he made his real money. In one IRS audit, the verdict went against him. The IRS is supposed to have declared that he owed them $150,000. Axel smiled as he paid.

In Oregon, Axel was up in Salmonberry country right next to government land. He had a key to open the gate, so, people say, he'd use it to go into the government property to hunt, any season, whether or not he had a license.

One day, the gate was open, so Axel went in, but, knowing the gate is not supposed to be left open, locked it behind him. He'd locked in the Game Warden, who was waiting for him when he brought down two elk. Axel just smiled.

The Ericksons were raised in the Manning and Buxton area, and young men from those small, northwestern towns have a reputation for being wild -- so wild, that when Axel's brother Emfred died young of a heart attack, someone explained it: "Everybody only has so many heartbeats in one lifetime. Emfred used them all up early."

Axel and his best friend, Dutch Schultz, logged out of Mountaindale, north and west of Portland. Dutch was just as ornery and bellicose and smart and tough as Axel. He was also even bigger. Dutch could sling a haulback block and four chokers over his shoulder and run up the hill with them. Axel would start a fight and Dutch would finish it. You had to take on both of them.

Axel and Dutch went to dances, like the one held every Saturday night in Scappoose in a big hall with wooden floors and a pot-bellied stove. Since every logging camp had its own still hidden away somewhere, the men at the dances were pretty well supplied with liquor.

One Saturday in Scappoose, Axel tried to dance on that wooden floor wearing his calk shoes ("cork boots"). That caused a real fuss as the spikes gouged up the floor. Soon men who had been drinking for much of the evening got up to join in, whether or not they knew what the fracas was all about. It wasn't long before the fight erupted into a full riot.

In the midst of the fighting, Axel picked up one man and set him right up on top of the glowing pot-bellied stove. It burned out the seat of his pants. Part of the restitution ordered was for Axel to buy him a new suit.

Axel and Dutch worked even harder than they played. For a while they were topping trees for Sherman. It was a huge outfit, but Sherman was having money troubles and couldn't pay in cash. Instead, Axel and Dutch were given forty acres of timber up on the hill.

One day Axel came in with a beat-up old logging truck and drove it up into the hill. That night some say he and Dutch "borrowed" a Caterpillar tractor somebody had left out to finish work with the next day.

There were also stories about the gravel loader they had. Rumors abound that Axel "borrowed" somebody's gravel loader, took it up into the hills above Gales Creek and buried it completely. Nobody ever found it. A long time afterwards, Axel is supposed to have dug it out and used it in his own logging operation.

However Dutch and Axel got their equipment, it wasn't long before they brought in a load of logs from their forty acres -- their first. But they weren't sentimental about it.

Axel wasn't sentimental about much of anything. He'd lost his mother when he was only twelve or thirteen. Not knowing what to do with the boy, his father took him with him up into the woods to get him work as a whistle punk.

On Axel's first day in the woods, a logger was accidently killed, and, to get the body out of sight, the men carried it up to the whistle punk's shed. They told the grieving boy to stay with the body and keep the flies off it. Axel was terrified, but he did what he was told. Somehow very few things seemed worth getting emotional about after that.

Axel bought a lumber retail company in Forest Grove and found out shortly afterwards that it wasn't making any money. He barked to his partner, "Sell the thing!"

"But what about the men working there?" his partner protested. "You can't just sell it."

"The hell I can't," Axel snorted.

Axel walked through a saw mill he'd just bought. He saw men who'd been working there for twenty or thirty years, but he was sure they were too old to work worth two cents. So he strode down the aisle firing one right after the other.

"I put the fear o' Christ in 'em!" Axel would say about the men who worked for him. His good workers respected him, but the lazy ones hated his guts. And the slow ones, he could pick them out in a hurry. He'd bellow, "Get your nose to the ground and your ass in the air and let's go!"

Early in their logging careers, Axel and Dutch worked one whole hot, dry summer to accumulate a large pile of logs. The pile caught fire, and there was no way to stop it. Axel sat down beside the smoldering mass and cried.

Logging was a hard life. It was hard work -- very hard, dangerous, and draining. Many men were killed in the woods. The wonder is that any of them survived it. Axel had rods in his sternum where the doctors had patched together his broken breast bone.

But in many ways, The Great Depression was even harder on the men than the hard life. They were tough. They had to be. Earl Nordgren, Jim Burns and Jerry Kemper knew Axel and Dutch in the late 1930's when the fighting team of Erickson and Schultz were in their 50's.

Jim Burns tells about Axel and a big German scaler, the man who measures the logs to determine the number of board feet the saw mill will pay the logger for. That scaler stood about six foot four inches and weighed a good 240 pounds. He always wore a jacket and a hat pulled down on his head, but he had a "long thumb." It is analogous to a butcher who presses the meat scale so he's weighing his thumb along with your meat. This scaler could rob a logger of almost a fifth of his total board feet when he scaled with that long thumb of his.

But Axel had scaled his own load before he took it in. When he saw what the guy was doing, he went back to his truck and grabbed a spare axle and went for him.

"I'll shorten that thumb of yours!" he yelled, and took off chasing the guy the length of the station and around the block.

The big German was running for his life, his coattails flapping out behind him and his hat barely clinging to the back of his head. Axel was gaining on him, until he hit the concrete sidewalk with his calk shoes. The metal spikes went skidding and sliding. Axel flailed and flung wildly trying to keep his balance, but he went sprawling. There was some expression on the scaler's face when he looked back over his shoulder at Axel going down.

But it wasn't merely for himself that Axel would fight for justice. Axel never stayed in the office of the saw mill he owned. He happened to be up in the crane overlooking where the log trucks were coming in when Earl brought in a load of logs. Then he saw Elton Beard come in and drive around to push in front in line.

Axel, nowhere near Beard's size, scrambled down from that crane and poked Beard in the chest. Beard blinked and backed up. It was Earl who explained that he had let Beard come around to take off his chains. Axel grunted and scowled. Beard was careful to see that Earl's logs were taken off first.

Axel figured all the angles. To have his logs scaled by the Forestry Department to determine fees, for example, he'd take whole logs, uncut, perhaps eighty feet in length. They measured the small end and did not count the tapering in figuring out board feet. The lower the number of board feet, the less Axel had to pay in fees.

But when he sold those same logs, he'd cut them to 32' or 40' lengths. That way the saw mill scaler had bigger ends to measure and Axel would gain as much as 40% in board footage to sell. The Forestry Department tried to stop him. They finally took away his branding hammer so he couldn't mark the newly cut parts of the logs. The story goes that Axel woke up old Van der Velden in the middle of the night to get him to forge a new branding hammer.

When Axel bid on a cutting job, he was skilled enough to estimate very nearly what he would make on it, so he often got the jobs he bid on. But then, if he wasn't making quite as much as he thought he ought to, it is rumored he'd move the boundary markers back a bit so there would be more timber to cut and more profit to be made.

Bargaining -- horse trading -- was the love of Axel's life. He'd use every trick he could think of, even when there wasn't much to be gained. While in San Francisco, Axel needed a taxi ride to the airport. He hustled from cab to cab in front of his hotel until he found a driver who agreed to take him there for $2.50. At the airport, Axel handed him a $10 tip.

Axel wore old clothes that were practically rags and drove an ancient truck filled with junk into Portland to sell the junk to a big corporation. He wandered around mumbling how he'd like to have a special piece of equipment he saw. The machinery was priced at $100,000. The company offered to let him pay in installments, but he refused.

"Nope," he told them, "I never buy anything on 'time.'"

He bargained them down to $80,000, then whipped out his checkbook and wrote out a draft for the entire amount.

Axel had a big, black Cadillac El Dorado. He'd drive it up into the woods on logging roads, and often where there were no roads at all. Somebody told him that that was no way to treat a nice car like that.

"That's how I made the money to get a nice car like that," Axel retorted.

Once Axel had a disagreement with somebody in business and the man refused to back down. Finally Axel asked, "Just how much are you worth? A million? Well, I'm worth eight million, so we'll do it my way." They did.

Jim Burns came upon Axel up in the woods personally dragging rock and filling in pot holes in the logging road. He was worth several million dollars by that time.

"Do you need to do that yourself?" Jim asked.

"You think I'm going to wait for somebody else? I've got to pay the bills every thirty days, you know."

Swede Ralston sold Axel a new Piper Apache airplane. It was the first of seventeen aircraft Axel bought from Swede, and he and the Ralstons became fast friends. Axel never did learn to fly any of the airplanes any more than he ever learned to swim a stroke. But he would stand in the bow of a boat yelling back, "Move faster. Faster!"

In his airplanes, he had no desire even to put his hand on the controls, but his son Jack became an excellent pilot.

When Jack was about 8 or 9 years old, he took a bite out of an apple and threw the rest away. Axel stormed over to pick up that apple and shove it back at him.

"You eat the whole thing," he ordered.

When Jack was a young man, Axel owned a saw mill and surrounding forest in Canada worth perhaps $500,000. But Axel was tired of the trips back and forth to Canada and offered to give it all to Jack.

"Nah," Jack is supposed to have refused, "I'll wait until you give me something worthwhile."

Dutch Schultz made a lot of money alongside Axel. Dutch never married and had no family of his own. When he died, he left everything to Axel's son Jack.

Jack is so slender and wiry, you think he could be a high school kid until you see his eyes. Those eyes are so intense and deep. Even when his net worth was said to be several hundred million dollars, Jack, too, loved to surprise people.

Jack went back East once to look at a sky crane Zikorski helicopter for his own logging operations. He was dressed down in hickory shirt and old pants. As he sat talking with the sales representative, he pulled out a can of tobacco and rolled his own cigarette.

"Are you sure you can afford one of these?" the salesman asked.

"Not one," Jack told him, "four." He pulled out his checkbook and wrote out a check for the whole amount -- millions.

Jack's mother was strict Catholic, and, once she'd had Jack, Axel complained about her so bitterly, his lawyer is supposed to have asked why Axel didn't just get a divorce.

"She wants $2,000,000 -- that could hurt a guy!" Axel snapped, though he was worth maybe ten times that much by that time. He never did divorce her.

The story goes that Axel was at a bar for some time watching a lovely blond waitress. When she came over to ask what he wanted, he told her he wanted her.

"Why don't you drop your husband and come home with me?" She did.

There is some dispute about what happened when her husband found out. Some say he stormed over to Axel's place to get her back and was not only bested in the argument but also had every piece of furniture in his apartment destroyed by a double-headed axe. Others say that was a different incident altogether.

For all his wildness, Axel had great courage and a cool head. He was once at the top of a spar tree when he lost his grip and fell. He tried to grab a cable on the way down, but it didn't slow him enough to keep him from hitting the ground hard.

Loggers rushed up and gathered around him, but he dragged himself to his feet, snarling at them to get away and leave him alone. Badly hurt, he went off into the woods. No one -- evidently not even Dutch -- knew where he went. It was three weeks later before he hobbled back into the logging site.

Axel was working another time in full gear, wearing heavy metal-spiked calk boots and a huge leather belt with axes, blades and ropes attached. He was bringing cut logs down to the mill pond when he fell in.

With that much weight on, he could only go directly to the bottom and stay there. It would take him far too long to shuck out of all that gear, even if he could do it under water. And, even if he could rid himself of all that heavy equipment and somehow get to the surface, there was little chance that he could swim between the logs without being crushed by the moving timber. There is almost no chance that anyone falling into a mill pond can escape drowning.

Axel did not try to get out of his gear, or to struggle to the shifting, menacing surface. He simply walked along the bottom until he could clamber out of the water at the pond's edge.

He was sixty eight when Swede Ralston's son Norm was working in the woods for him, setting chokers. Lightning had started a series of small fires the night before, and Axel came walking up to Norm carrying shovels and picks and a five gallon tank of water in each arm.

"Here, take these and follow me."

Norm struggled into the backpack of water, grabbed up the digging gear and hurried after him, looking for any small fires the lightning had started deeper in the woods. But Axel was already far up the hill and Norm, though in peak condition, was hard put to keep up. After a hard trek into the woods, Axel looked back.

"You look pretty tired. Stop here and put out that stump," Axel ordered the nineteen year old Norm, and he hoisted his own gear and continued on up over the rise.

Fishing one time deep in the back woods in Canada, Axel lazed in the boat, chatting with the pilot who'd flown him up from Oregon. He really didn't want that time to end until, as always, he was ready for the next thing. The two men rowed back to the plane, stowed their gear and took off. But as they started circling over the lake, Axel yelled, "Hey, go back! There's a moose down there."

"Are you sure?" the pilot asked.

Axel, who had hunted many times in Canada and Alaska, snapped, "Hell, of course, I'm sure. I know what a moose looks like."

They circled and landed. The two men hiked out after that moose and got him. Axel was delighted with the kill. They cut it up and were hauling the meat back to the plane when Axel straightened, then bent over, dead white and sweating. His hand went to his chest as he stumbled forward.

He dropped the moose meat and lay down beside the trail.

The pilot stood nearby trying to decide whether to hike on out and go for help, or whether he could load Axel over his shoulder and try to pack him back to the plane. Forget the moose meat.

But when the chest pain subsided, Axel got up and threw his load of meat over his shoulder.

"Come on," he ordered. The pilot hoisted his own load of meat and followed.

In the middle of a funeral for one old logger, Axel stood up and cried, "Well, Maynard, may you go where the trees are tall and there ain't any inspectors." We hope something of the same for Dutch and for you, Axel Erickson.

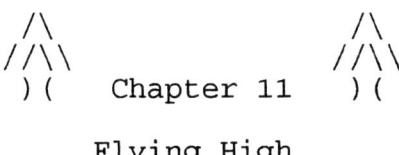

Chapter 11

Flying High

There's the story of the old farmer and his wife at the fair who were fascinated by the young man's flying machine.

"How much for a ride, Sonny?" the farmer asked.

But the price the aviator named was more money than the two old people saw in a summer's work. They started to hobble away.

"Say," the aviator called after them, "it's been real slow today. Tell you what, I'll take you up and maybe that will get other people wanting to go, too."

"How much?" they asked, suspicious.

"Free -- on one condition. If you come through without hollering or crying out, I won't charge you anything. But if you say a word, you pay full fare, okay?"

Now "free" was the kind of economics the old farmer liked best, so he helped his wife up and then climbed in beside her in the back seat. "Take 'er up, Sonny," he hollered, and then he shut his mouth tight.

The pilot took off with a flourish and buzzed the field and the fair. When he looked back, the old people were smiling with their eyes open and their mouths closed.

"I'll get him," the pilot said to himself. He climbed and did a loop, and then a stall, and finally a barrel roll until he was starting to get a little airsick himself. Shrugging, he flew back to land at the fairground.

"Say, old timer, you did it. I didn't hear a peep out of you," he remarked as he helped the farmer out of the plane.

"Well, now, Sonny," the old man quavered, "you almost had me there once, though, when the old lady fell out."

In the early days of flying, before the invention of sophisticated instruments, a pilot flying into heavy clouds often took along a cat so he could tell which way was "up." A cat will lean to keep upright.

Swede Ralston was one of those intrepid pioneers of aviation. He remembers when he was too poor to be able to afford a watch, but he scraped together enough for flight lessons. He would "buzz" his girlfriend's house low enough for her to be able to see him tap his wrist. She'd run into the house while he circled and back out to signal up what time it was, so he'd know when to head to the airfield to turn in the rented plane.

The tiny airstrip at the Flying M Ranch was short on amenities, but the pilots of the small planes and helicopters that landed there loved the rugged outdoor setting of Bryce Mitchell's remote dude ranch.

Bryce hosted lumberjack shows on the Flying M, often in conjunction with the Portland Rose Festival in early summer. Despite the rain, one emcee kept the crowd happy with his deep voice and lively patter. But when his stint at the microphone was over, he plopped wearily into a folding chair.

"I run the show at my place, and the pros are done in one and a half or two hours, instead of hours and hours the way it is here with all the amateurs. The trouble is, this once-a-year pretending you're young -- it can kill ya!"

Bryce's daughter Beth had loved animals since she was a tyke nursing robins with broken wings, trail horses traumatized by inexpert riders, and raccoons with sore paws. So when she heard that a hurt cougar was scheduled to be destroyed, she couldn't bear not to try to help it. She'd already sent so many animals back to the wild in good health. How much harder could it be to nurse a cougar?

How much harder, indeed.

But Beth, determined to try to save that cat, took on the cougar's care. At first it snarled and snapped, and then, gradually, it began to accept and finally to trust Beth and her family.

And then it came to love her. It even wanted to sleep in bed with her sometimes. Beth wore it draped around her shoulders as she drove.

After a year the cougar was well enough to be sent back into the wild. But it had grown much too friendly with too many local people to be released anywhere near the Flying M Ranch.

A friend with a small plane agreed to take it up to the wilderness in Washington state where the cougar would be free, and far from the Mitchells. Ed Chadwick helped Beth wrestle the large dog kennel into the back seat of his plane, then watched her coax the tranquilized cat into the kennel. She bolted the door, draped a blanket over the kennel and signalled for Ed to get in.

He'd watched the sheer power in that sleek form, and, though he didn't say anything, it was with some trepidation that he climbed in beside her and took off over the trees at the end of the Flying M runway. He climbed and banked, heading north and east. They were nearing Forest Grove when sounds behind him made the hairs on the back of Ed's neck begin to rise.

Wide-eyed, Beth twisted to reach back and lift a corner of the blanket.

"Oh, boy," she whispered, and Ed's scalp stood tall. He shot a horrified look back over his shoulder.

The cougar had gnawed through the door of the kennel and was reaching out a paw!

"Stop!" Beth commanded, and, for a moment, the cougar stopped. But only for the moment, and then it worked again at the hole in that door, gnawing and pawing.

"Stop!" Beth spoke again, but she was already pulling out the pistol she'd brought along -- just in case.

At each command, the cougar hesitated, then resumed its work toward freedom. But that freedom meant confinement with Beth and Ed in a four-seater airplane flying a thousand feet in the air over Forest Grove.

"Uh-oh, Beth, can't you control that thing?" Ed groaned, while she tried to find a way to secure the cage with one hand while aiming the pistol at the cougar with the other.

Even the few minutes it would take Ed to get back to the Flying M were more than he wanted to spend cooped up with a cougar within inches of the back of his head in an airplane only he could pilot.

"Banks," Ed said. The small town and strip were dead ahead. Please don't let that "dead" ahead be literal, flashed through Ed's mind before he wiped out all thoughts other than getting this flying wild animal cage on the ground ASAP.

They landed. Ed taxied to one side, braked to a rocking halt, switched off the engine, leaped out of the plane, and ran.

"Come on, Beth!" he called.

But Beth Mitchell couldn't shoot that cougar any more than she'd been able to let someone else destroy it when it was hurt.

Beth stayed in the plane, talking and cooing, singing and talking, and gradually, with Ed gone and Beth's beloved voice droning, calming, reassuring, the cougar settled down.

Ed called the Flying M to tell them what had happened. Beth stayed with the cougar until they could bring a truck to take that cat to Washington to release it.

One offer of flying cougar transport was more than enough for Ed Chadwick, though he had to admit that the moving job hadn't been boring.

"Can you imagine how discouraging it was to ride horseback through such country," Earl sighed as we drove through the desolate hills between Lake Hemet and Palm Desert, California. Earl was not a noted horseman. His only story on himself was the time he came riding down the hill above Dr. Vern Jackson's farm west of Forest Grove. The saddle slipped, and Earl ending up sitting on the poor horse's neck.

"Imagine, riding day after day, in country like this with no trees, no shade, no water and with tomorrow having no more promise than today? But at least they'd sleep good," he scowled. "Though it's hard to find the soft side of a rock."

Not that Earl Nordgren hadn't done his share of roughing it. As though his years in the woods weren't rugged enough, as his son Al grew up, Earl took the bright-eyed boy on hunting trips to Canada. They stayed in cabins with saw blades set on edge in the narrow window sills to keep the bears from reaching or climbing in.

Al loved it! He talked his dad into taking him back year after year. On one trip into the deep woods they found a moose that had been mauled several days before. It was stuffed in a hole and covered up with leaves. There were grizzly tracks all around that stash big enough Earl put both hands in side by side with his palms flat.

They knew that grizzly bears sometimes half-bury their kill and cover it with leaves for a few days to a week and let it decay until it was ripe for eating. So they took their guns and climbed a nearby tree and settled in to wait in hopes of seeing the grizzly return. They waited hours, but the bear didn't come.

One morning one of the hunting party came running into the cabin yelling, "There's the biggest moose you've ever seen on that island." The other hunters poohed his enthusiasm, but they did get their guns and go out to see for themselves.

There was a moose on the island, and it was enormous. The men decided that they would all shoot at the same time, on signal, so each would have a chance before the others' shots scared the moose off.

Al and Earl didn't have the range to make it across the expanse of water. But Forest Grove neighbor Bud Pietsch had a much larger gun. Bud got that moose with his first shot. Twelve year old Al watched it all with enormous eyes.

The next hunting season, they were rowing in a small boat and saw a moose swimming in the lake trying to get away from them. But Earl rowed hard and caught up. Al was so excited, he jumped out of the boat into the water with the moose.

Al also shared his dad's love of flying and became an excellent pilot, often flying his seaplane up into Canada or Alaska to hunt or fish.

Al and a friend spotted a great fishing cove from the air, and set the Goose down on its pontoons in the middle of the cove. They anchored the Goose, and set out in their small boat with their fishing gear.

It wasn't for quite a while as they sat throwing in their lines and pulling up big ones, that Al realized his floatplane had been slowly shifting in large, lazy circles. Sitting up a little to watch, he measured its movement against trees on the shore. The circles that were getting smaller and smaller.

"Whirlpool?" Al said, and his friend looked up. "Let's get out of here!"

By the time the two were loaded and ready to take off, the eddy current was strong enough that the Goose had tough going, straining to lift off the surface of the water.

Ron set his Goose down in a high Alaska river and put up camp along the bank. Eskimo men, paddling in from ocean fishing, saw his campfire and stopped at the edge of the water.

"Come on in," Ron called, "and get warm here by the fire."

But the men just looked at him, astonished that he would suggest that Eskimo men, in their furs, in the summer time, could possibly be cold.

"Well," Ron offered, "come in and have a cold beer."

Al was flying up in the back country of Canada when his friend spotted something on the rocks beside a wide river.

"Hey, go back! Walruses!"

They flew back and buzzed the rocks, but what was sunning wasn't walruses, it was naked Eskimo women, who waved and smiled as the plane flew on.

Earl Nordgren had hobbled on his fused left hip for a number of years when he went to a logger's meeting in Eugene and saw a helicopter set up on the convention floor. He'd flown fixed-wing planes, but when he saw that chopper, he fell in love. Never mind that the fixed-wing flight instructors call helicopter pilots "rotorheads" and "not the sharpest knives in the drawer." That chopper looked like fun.

The trouble was, with his hip fused straight, Earl couldn't sit inside and manipulate the pedals.

"When you want to do something, just do it," is Earl's motto.

He talked to his orthopedist and they scheduled joint replacement surgery to give him better flexion at the hip.

As soon as he'd learned to walk and maneuver again, Earl took helicopter lessons. His son, Al, took lessons, too, and another of the long-experienced pilots from Hillsboro. The lessons came hard to Earl, but the other two breezed through, confident and skillful. Until the testing. Earl had studied long and hard. He passed.

The other two did, also, on their second try.

Earl and Al were flying Earl's helicopter above the Tualatin Valley in Oregon when fog settled in so thick that all they could do was to ride along just above the telephone poles. That way they were pretty sure there wouldn't be anything in their path.

It reminded them of the stories Alaskan helicopter pilots tell about fog on the Arctic coast being so dense, they have to fly so low over the ground that they pray that a moose or an elk won't stand up too suddenly in their way.

Old timers can't resist pointing out to a helicopter owner with mechanical problems, that a horse may be slow, but it is dependable. And a horse doesn't eat all that much.

With the main rotor of a helicopter spinning in one direction above the cab, a tail rotor is needed to counteract that one-way motion. The tail rotor is small and whirls so fast it appears as a faint gray blur, all but invisible. People have been decapitated when they get too close to a spinning tail rotor.

Al knows that. But even with as much experience as he has had in and around helicopters, he was still very nearly hurt as he inspected a chopper's paint coat. He kept working toward the back of the bird without even realizing the tail rotor was going until a friend yanked on his arm.

"That's close enough, Al," he said.

Al swallowed, and was glad he still could. How easy it is to get into trouble without even realizing you're in danger.

A visiting friend wanted Earl to fly his helicopter up the Columbia River to see the new Trojan Nuclear Power Plant northwest of Portland.

"Why not fly over it?" Earl suggested, and called agency after agency until someone in Portland finally granted him permission to fly directly over the stack. Only nobody called Trojan to let them know.

Earl flew over the stack, then landed his helicopter in the parking lot beyond the Trojan Visitors' Center. He was startled to see patrol cars roaring toward them. It took a while for the security forces to confirm that Earl really had gotten permission to fly over and that he wasn't some anti-nuclear protestor threatening to toss a bomb into the stack.

The chimney route got old for Santa one year and he asked to arrive in Forest Grove in Earl's helicopter. Poor Santa turned out not to be much a flyer without his reindeer, however. He was so glad to be on the ground, he got out a little before they were.

When a friend of Al's was married at Pacific University in Forest Grove, instead of hiring a limousine, he asked Al to pick up the newlyweds in the helicopter and fly them away from the ceremony on the college campus. The new bride tossed her bouquet from fifty feet up and rising.

Earl and Al got to be familiar sights over Forest Grove and the surrounding area. Earl sometimes dropped in on friends out in the country. The little granddaughter of one couple Earl visited would smile and point up. "Here comes Earl and his 'Hello-copter.'"

One time Earl landed in the bank parking lot to take a vice-president for a ride at the instigation of her fellow-bank workers. It was Margaret Johnston's first flight, and she wasn't so sure. But she was game. She loved it, once they got airborne.

Pioneer aviator Swede Ralston's son Norm is himself an excellent pilot. Norm and Earl flew the Nordgren helicopter into the Grand Canyon before laws were enacted to stop the practice for environmental and safety reasons. There was a cable stretched across the canyon that had been used for years to haul bird guano out of a cave near the floor and up to the rim for transport to the fertilizer factory. To hit that without realizing it was there would have smashed the rotor blades and hurled them to their deaths.

Norm was so skilled he flew them down holes and between pinnacles where there wasn't a foot and a half to spare on either side of the rotor blades. They landed on top of a pinnacle and got out to cup their hands around their mouths and holler "Yahoo!" up and down the canyons, grinning as their greeting was echoed back at them.

Norm and Earl hadn't heard yet of the pilot from Hillsboro who had engine trouble as he flew over Grand Canyon-like territory in the West. He was barely able to auto-rotate the stricken chopper to land atop a slender pinnacle.

Heaving a sigh of relief at his miraculous landing, the pilot started to get out of the chopper when he felt the whole pinnacle shiver under him. The column collapsed, crushing chopper and pilot. He'd come so close to saving himself from one disaster only to die in the second.

In the desert near Flagstaff, Arizona, Earl and Norm got permission to fly the helicopter into a meteor crater so deep and wide that a boulder on its far rim looks like a good sized stone, but is actually the size of a small house. At the Visitors' Center, large chunks of nickel are on display that were brought up from the meteor that created the crater. The meteor hit so hard it buried itself almost a mile deeper than what looks to be the bottom of the crater. When it fell, the meteor crash scattered dust for miles around. Indian legends tell about the fire in the sky and sparks broader and higher than men could believe.

Charles Pritchard kept trying to get his father to fly with him in his early Piper Cub. But his dad was bound to keep "both feet on mother earth."

Charlie talked him into going out to the airport with him, and even talked him into walking with him up to the Cub to look inside.

Charlie had fitted the floor in front of the passenger seat with grass sod. Realizing he'd been had, James got in and let his son taxi and take off. But they were hardly "above the pigeons" before James said, "That's enough," and got him to land. James never would go again. Mother Earth was made for men. The sky was made for... James only shrugged. He wouldn't say.

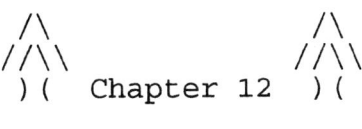

Chapter 12

FIRE!

There is an old story about two hikers fleeing from a bear. One pants to the other, "You can't outrun a bear."

The other answers, "I only have to outrun you."

Where the bear would be content for a while with one of the hikers, a fire can outrun and consume them both, and anyone or anything else in its path.

When a fire roars uphill there is no way to stop it. Heat rises. By creating its own very hot gases, a fire can create its own 30 to 40 MPH wind. From its own internal mechanisms, a fire can build such tremendous force, it can whip the top right off a tree four feet in diameter and throw it up the hill.

The Nordgrens sometimes used a slashburn to clear the debris from land they had just logged. The men patrolled carefully, throwing dirt on tongues of fire that threatened to go beyond the cleared area. But one time the fire got away from them. It jumped the road and started running up the hill miles an hour.

A fire in timber country can destroy thousands of acres of prime forest, to say nothing of houses and buildings -- and life. Fire engines, tank trucks and firemen rushed out to the woods.

Fire fighters, carrying heavy five-gallon pumper packs on their backs, used hoes and shovels to clear a break around the blaze. A bared space often stops a fire by depriving it of fuel. But an errant gust of wind can help a fire jump across that cleared line into new timber.

The Nordgrens kept a water truck by their donkey all night, hoping to save as much of their equipment as they could.

Hours after the fire had started, Fred Nordgren and Joe Zberg hiked up over the hills to see how it was burning. They crested a rise and looked down into an inferno. The fire roared uphill at them.

Both turned and fled, barely able to run fast enough to save their own lives.

It was that day and night and the next day before the fire was finally under control.

Loggers, especially those who owned their own land, could never relax after an exhausting day of work in the woods.

Earl seldom left home in the summer. Evening after evening, he sat looking out at the line of hills west of Forest Grove where his timber grew, always alert for wisps of smoke. A rising column of smoke could mean a fire that would wipe out everything they had been working for for years -- gone in a matter of hours.

If a tiny smoldering fire can be detected right away, it can be finished off merely by stamping on it with boots -- a far cry from what is needed if it gets to burst into flame and feed itself in voracious frenzy on the timber.

One hot, dry day, the Nordgren's haulback line got caught under two spikes they were using to hold down a guy wire. As the donkey engine pulled on the line, friction heated the spikes until finally the glowing ends split off and were thrown, red-hot, into the brush. The loggers rushed over to stamp out the fires.

Once even a glass water jug started a fire as the sunlight was concentrated through its sides, like a magnifying glass on dry tinder. Luckily, the men smelled the fire right away. They ran over and put the fire out with the water in the jug.

In dry seasons, the Forestry Department would shut down all logging operations if the humidity level was too low. Usually that was a good indicator of the degree of danger. But Rutherford remembers when all logging was shut down once because of low humidity, while he had a foot of snow on his landings.

In dry season, loggers worked the "hoot owl" shift, from daybreak until noon, and then quit, because the danger of fire was just too great in a hot afternoon. A watchman stayed on site for at least two hours after the others had gone, just to be sure there was no smoldering fire started from friction or from a spark from one of the pieces of equipment.

In his later years, Alfred agreed to do the firewatch hours after Earl's hoot owl shifts. It can be so lonely in the woods, at times Alfred's attention would be caught by anything that moved. When he saw a bee circling and buzzing at a distance, Alfred watched it drone on in its haphazard way, until it gradually came closer and closer, landed right on the end of his nose -- and stung him!

On another fire watch, Hildur sat with him for company. As Alfred looked off scanning to one side, Hildur scanned the other. Her eyes fell on a tall dead tree. The snag's top was bent far over.

"That snag is going to fall some day," she said, and as Alfred looked over, it did fall, right then, as though her words had precipitated the event -- or the snag had simply been waiting for an audience.

Snags are dead, dried-up trees, mostly just trunks left. A lot of charred snags are left after a forest fire, many still containing good wood that can be harvested, before worms and bugs chew it to sieves.

In a fire, a snag acts like a chimney. Sparks are hurled into the air as though shot from a cannon, landing on other trees and starting more fires.

The old Indian and Boy Scout trick of starting a fire by rubbing two dry sticks together can be recreated unintentionally by the loggers merely by dragging one dry log over another. The infamous Tillamook Burn that was so wide-spread and so devastating was probably started by friction sparks.

1933 was such a dry year that the state authorities had just about decided to halt all logging because of the enormous fire hazard.

Up in the woods one outfit was working under their spar tree, hauling a heavy cedar log. They pulled it over another log and the friction of dry bark against dry bark was enough to set sparks that burst into flames that spread so fast the desperate men could not contain them.

To make matters worse, another fire started several miles to the west, and no one could stop it either. Quickly, the two fires joined up and roared to devastating life that threatened the Northwest.

And even worse, someone started a third fire north of the big two. They hadn't realized yet how devastating the Tillamook Burn really was. It wasn't unheard of in the Depression years for an arson fire to be started just so the men would have work putting it out.

There were several other enormous fires during the 1930's. Three different times during those terrible burns, fire came up right to edge of the Nordgrens' property at the east rim of the Tillamook Burn area -- and stopped of itself.

Why? No one understood why. A difference in moisture content in the soil or the water content of the wood that kept the Nordgren trees resistant to the flames? A difference in the prevailing winds that refused to drive the fire forward? They never knew why for sure, but they stood in awe each time, helplessly watching as the fire's forward progress stopped by itself. They knew there was nothing they could do if it hadn't stopped on its own.

In the early days, the steam donkey boiler was fed by an open fire. Stray sparks flipping out into the dry brush, sawdust or timber were such a hazard that crippled old men or boys were hired as "spark chasers." Besides stamping out fires before they could take hold, their other job was to keep the water tanks filled.

Jerry Kemper's first job in the woods was spark chaser. He was sixteen. He lied about his age. You weren't supposed to start at that time until you were eighteen.

Art Vaandering, one of the best loggers Earl ever knew, was hauling a crane with a loader one very dry summer day. The loader back-fired and set the motor on fire. Art knew that the whole stand of timber could go up faster than they could get out of the way, but he had nothing to put the fire out with.

When trouble happened, Art's adrenalin got pumping so, no matter what it was, he could handle it. He threw that hood open and scooped up dust and dirt and threw it on the burning motor. He wouldn't stop until he got that fire out, though his hands were raw and bleeding.

Sawdust burns on and on and on once it gets started. It keeps on smoldering low so you can never be sure you've really put it out completely.

A Seventh Day Adventist man ran the saw mill way back on Earl's property. One weekend someone saw smoke and called to warn Earl, who called the mill man. But the Adventist's religion forbade him to work on Saturdays.

"This is an emergency!" Earl hollered.

Earl jumped into his car to take off for the woods. He and the mill operator used the big caterpillar tractor to bulldoze the sawdust pile flat, scattering it. Then they wet it all down. They dozed it again, and wet and dozed and wet the pile again and again all through Saturday night and into Sunday morning.

There wasn't much rest on that week's Sabbath.

"No rest for the wicked," Earl would say, knowing he was one who seldom had a day off.

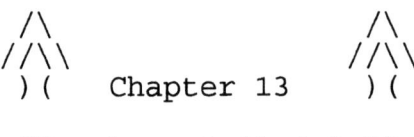

Chapter 13

"Tough as Boiled Owl"

Earl scowled and shook his hand, then went right back to lifting his end of the old chest of drawers. He'd gotten a wood splinter deep into the pad of his index finger, but he shrugged.

"No time to go digging at it with a needle. It'll rot out," he said. His attitude toward pain is not unusual among loggers.

Nearly everybody who worked the woods got broken up. Look at any old logger and see how deformed his hands are. There were a lot of broken bones, but the men just kept working.

Before Earl's first time topping a spar tree, when he was about eighteen years old, he was working for his father in the woods. The cable around a stack of logs beside the road was not quite right. It needed to be adjusted so the cold deck could be moved.

Hopping onto the logs, Earl ran across to the cable, but he tripped and fell. Instinctively he thrust his arms forward to catch himself. Both hands were bent back at the wrists so that the backs of the hands were jammed flat against the forearms.

They hurt some. They hurt a lot. But he toughed it out and stayed on the job.

Later that same afternoon, Alfred needed someone to pull a cable through the pulley at the top of the crotch line pole so the cold decks could be lifted and loaded. There was no other high climber, so Earl gritted his teeth and roped up the sixty foot pole to pull the line through.

Fifteen years later Earl was having a physical by Dr. Forrest Bump.

"Any complaints?" Bump asked.

"Well, yeah, these wrists hurt some, particularly in the rainy weather."

"Let's X-ray them and see what's wrong." The x-rays showed both wrists had been fractured and distorted and allowed to heal however they could. And through both wrists were cracks where the bones had never healed at all.

"Pretty tough Swede," Dr. Bump said, shaking his head.

Earl wasn't the only son of an immigrant working in the Northwest woods. Many of the early loggers were immigrants, or the children of immigrants, particularly Scandinavians or Switzers, and that fact led to some wonderful mixing of traditions -- and language.

One tradition that was quickly adopted was that whenever two Norwegians get together, it's time to eat.

An immigrant who worked occasionally for Alfred got frustrated with one of the Nordgren horses and finally swore at it in fluent Norwegian.

"There," the man said, calming down afterwards, "you've not heard THAT before."

Joe Zberg, (whose sister married Joe Zgraggen), used to tell red-headed people, "I hate to have you die -- you have such pretty hair."

Joe was heard giving somebody directions including, "...turn the corner around twice."

Zberg had been born and raised in Switzerland, where the farmers' land was so steep it was impossible to use horses or even sleds. They had to carry their hay, and sometimes even soil to make a garden, on their own backs. The men would haul enormous bundles to the barn, which had only normal sized doors, not extra wide ones as in this country.

Before he came to this country, Joe knew a poor family desperately in need of wood to burn to warm their children. They stole a tree from the state forest and buried it in the manure pile. It was the only way to hide it from the authorities.

When Joe Zberg came to this country he worked helping Alfred and then Earl Nordgren with the logging. He built a cabin way back in the forest. One day a skunk visited Joe and left his calling card. Many days later, Joe dressed up in his suit and went to town. "Somebody sniff air and growl, 'Smells like skunk in here,'" Joe told Earl. "I slink out. I know who skunk was."

One of Joe's partners got sick after a heavy night of drinking. As Joe explained it, "He threw himself out and I laugh at him almost."

One time Earl and Joe drove to Squaw Valley to see the Winter Olympics. After the Olympics, they drove on to Reno to gamble a little. Joe watched an old lady who was playing a one-armed bandit, long before the days of electronic buttons. She was so disappointed, not wanting to stop, but her arm was too tired to pull the lever down. So Joe sat down beside her and pulled down the lever as she fed in the nickels. She won just enough to keep them both at it all night long.

Swedes were noted for their unromantic view of the world. Earl had been told that an uncle of his back in the old country was supposed to have growled at his wife when she mooned for affection: "I told you when we married that I loved you. If it ever changes, I'll let you know."

But even Earl was taken aback at Joe Zberg's story of an unsentimental family he had known in Switzerland. Joe's friend was mountain climbing with his father and older brothers when the father fell down a chasm and was killed. The sons stood at the edge looking down.

Finally one brother said, "I guess we better climb down there and get his watch. Otherwise we'll have to come back."

Joe Zberg was raised a Catholic, but he didn't practice the religion so much as the basic philosophy. "I just believe in doing what's right," he said, and lived.

Gwen, a Catholic born in Ireland, grew up terrified to cross the street in front of their family home. There was a sign posted saying, Pedestrian X-ing, and Gwen was afraid it said Protestants, so she would only cross when her dad was there to hold her hand.

Earl's father Alfred named his first-born son Manfred, (which Fred quickly shortened). The family story goes that as a newborn, Fred cried and howled the first days of his life. Nobody could stop him from crying. Aunt Selma, the midwife, finally gave him a shot of whiskey and baby Fred slept for three days.

One time when Fred was a kid he brought home a big, bright belt buckle something like a cowboy buckle, but Alfred and Hildur wouldn't let him wear it. He was mad. He thought it was kind of pretty.

The Nordgren boys grew up working in the woods with their dad. Fred was a big man, powerfully built and strong as an ox. Paul Bunyan-like, Fred used to kid he'd found one special peeler tree on the Oregon coast that would be a whole summer's work to fell and cut up and haul out.

Alfhild and Charlie Zumwalt's twins got along so well together growing up they even shared a checking account. But Earl didn't do that with Fred. Fred would write checks, all right, but probably not make many deposits.

"You should have said so before you spoke," he'd tell younger brother Earl. Or, "You don't think -- you just think you think." When Earl started looking at the Bible, Fred told him he must be going rat-trap. "You don't have to believe it, but it's a lie."

"My brother joined the Navy in peacetime to get out of the home place, I think," Earl says. "Fred thought the country life was boring. He used to say, 'I want to be where the people are thicker.'" With his good looks and ready wit and smile, Fred was always popular with a crowd.

Fred had a heart attack when he was only in his fifties. His small son Tim was with him. Terrified, the boy ran from house to house trying to get someone to help until he finally found a woman who understood and called an ambulance. They got Fred to the hospital in Seattle and it looked as though he might pull through. But he was told he had to lie perfectly still and let his heart heal.

"That's a bunch of B.S.," Fred growled. That night he got up alone to go to the bathroom. His heart "blew out," and he died.

"Some people you can talk to," the doctor told the family, "and some you can't."

Roughhouse Dickson was another great big man who made up his own mind. Roughhouse went to a logging carnival on the Oregon coast at Seaside. He watched the sawing races and springboard races and pole climbing races. But the longer he watched, the more his face set in a sneer.

Many logging shows use a ringer in the crowd -- sometimes dressed in a tuxedo -- who will holler loudly that the regular contestants aren't doing a very good job climbing. The announcer will pretend to talk to the "drunk" and warn him not to try to climb the pole. But the guy will climb anyway, much to the audience's concern.

At the top of the pole, the ringer will, unseen by the crowd, attach the harness he is wearing under his jacket -- or dinner jacket -- to one of the guy wires with a hook. As the crowd oohs and aahs, the guy will pretend to be so drunk he loses his balance at the top of the hundred foot pole and falls. The crowd gasps. But in reality he slides safely to the ground along the guy wire.

Most of the people who heard Roughhouse yell out must have thought it was a set-up, but they gathered around anyway. The ringer's performance is always a thrill, even when you're sure it isn't what it seems to be.

"I can climb that there spar tree," Roughhouse challenged, "even without climbers." And up he went, hand over hand up the guy wire. At about sixty feet above the ground and the gathering crowd he lost his grip and fell. He landed on the top of somebody's car and made a long, deep dent.

They scooped him up and took him to the hospital unconscious. When he woke up in bed, he got up and walked away.

Alfred had not been walking long on his artificial leg, but, despite the rough terrain, he had to find a way to get around in his woods. He carried a cane, which worked pretty well, until he got to a creek. The only way to get to the other side was to maneuver his way along a tree trunk that had fallen across, spanning from bank to bank.

"God, I'm scared," he whispered, but he gritted his teeth and hiked up and across.

Earl was out walking beside the log pond by the river in his slick-soled Oxford shoes. Though he wasn't wearing his calk boots, he decided to check the raft of logs out in the water. He had stepped out across the two skinny poles to the raft many times before without a problem.

But it had been a while since he'd crossed here. Only when he'd already started did he realize the poles were now algae-covered and so slippery he could barely keep his balance.

Once he'd started, though, there was no way to turn back. He had to keep going on over to the raft. If he fell into the river, the current would carry him under the raft and he'd never be able to come up for air. So it was a matter of life or death balancing all the way out. And then he had to do it all over again to get back to the bank.

"I did some dancing."

Jim Burns remembered eating in the chow tent deep in the woods when the men were suddenly aware of the shadow of a tree falling down over the tent -- and them.

"It was the biggest damn scramble you've ever seen in your life," Jim chuckled.

They had a new gas stove in camp, and Stew Petit was pretty sure he could figure out how it worked. He got it set and lit a match, and the new stove blew up in his face. All his facial hair was burned off, including his eyebrows. Later, when he'd tell the story, he'd smile innocently and add, "So then I read the directions."

Stew went to blast a stump one time. He didn't want to get his brand new watch dirty, so he took it off his wrist and set it on the stump while he dug the ground to make a hole to put the blasting powder in.

He set the powder and lit it and just as the stump blew into smithereens he remembered the watch he'd set on that soon-to-be-sawdust stump.

"It blew clean away. There wasn't nothing left."

When a tree dies in the woods, it will usually keel over and lie as a rotting log on the forest floor. But some dead trees remain upright, limbless and leafless hulks, standing with no visible means of support. Many snags still contain good wood that can be sawed and planed into building material. But when snags are cut, there is no predicting which way they will fall.

Earl Nordgren and Art Vaandering weren't able to log one inclement winter, so they went up into Earl's woods to harvest snags. Wanting to be sure they didn't hurt each other when their snag fell, they were working separately, often more than five hundred feet apart.

After a while, Art moved completely out of sight, working his section of the woods.

Suddenly Earl heard him hollering, shrieking, really. Earl was sure from the sound of his voice that he was badly hurt.

Throwing down his equipment, Earl took off running to help him. He got to him just in time to see Art finish killing a bear with his axe. Art had disturbed it as it slept in the hollow of a log and it had showed with its claws just how much it resented the intrusion.

Earl and Art packed that bear back down off the hill and Art had the hide mounted. Not many men can say they've killed a bear with an axe.

The next time Art and Earl went up there to fell snags, Art got to hollering again, and Earl wondered if he'd killed another bear. But this time he was hurt.

A huge load of bark had fallen out of the tree Art was working on. The bark fell on Art's legs, bruising them terribly.

Earl knew about the danger of chunks of wood falling out of trees. He once let go of his end of a large, 2-man power saw to walk around the trunk to check how close other men were as they cut. While he was away, a enormous chunk fell, burying itself deep in the ground between the handle bars Earl had just been holding onto.

Earl tried to pack Art down the hill, but Vaandering weighed better than 220 pounds and Earl couldn't pack him far. So Art hobbled out on one leg. He was laid up for more than a month.

Art Vaandering was such an ambitious, hard-working logger, he used to go out on his own while his crew was on their lunch break. Once he went without his hard hat.

He'd cut his tree through and ducked behind a fallen log for cover, but he hadn't heard it fall yet. So he lifted his head up to see what was wrong, just when the tree came down. A chunk of wood split out, sailing through the air to hit him square on the head, killing him. If he'd had his hard-hat on, he might have gotten off with a headache. Art wasn't a lot more than thirty years old. He left a wife and three little kids.

Earl and the other loggers were devastated at the news of Art's death, but loggers cope with many tragedies.

They all hated to be the ones to tell a woman her man had been hurt in the woods. Earl still shudders remembering the time he was asked to tell a neighbor that her husband had been killed. Two or three other men went with him, and she knew as soon as she saw them. Her face went ashen.

Carl Johnson was a log cutter who worked for Earl until a month before his death. He was cutting for someone else on a steep slope and then went further down to cut another log. The upper log started to roll. There wasn't time for Carl to scramble out of its way and he was crushed.

Ernie M. was a welder who drove from one logging site to another to work on equipment. He carried his tools in the back of a beat-up old flatbed truck with no brakes and no clutch. He would rev the motor until he heard the change in pitch of the engine sound, then quickly shift gears.

Earl and his father, driving to work, saw him as he started up a steep grade ahead of them. Earl looked over at Alfred when they recognized Ernie's truck. Alfred didn't say anything. What was there to say about a man who had only a few weeks before had lost both of his small twin sons? He'd taken them with him to the saw mill. While he was working, the boys went down to the mill pond. Both fell in and were drowned.

Sighing, Earl slowed as he realized Ernie's truck was having a hard time getting up the 30% grade. The Nordgrens could see that Ernie's truck was losing power and slowing. Ernie must have tried to shift gears quickly enough to catch a lower gear while he still had momentum, but he missed. He started rolling backwards.

Rather than risk rolling all the way down, Ernie steered into a bank. The truck tipped over, and his tools were strewn all over the road and halfway down the hill.

Ernie hauled himself out of the truck and slumped over to sit at the side of the road. His expression told Earl he really didn't care whether he lived or died.

The Nordgrens parked behind Ernie's truck and gathered up the scattered tools and helped him tip the rig back up on its wheels, then pushed him up the hill and went to work. As Ernie got the tools in his hands, he worked, robot-like, but he kept working.

Alfred was asked many times how he could stand the pain when he lost his leg. He could only shake his head. What other choice did he have but to stand it?

Alfred and his sons had been having trouble with their gas donkey. It was just not working right. The donkey had a heavy crank about four or five feet long with a foot-and-a-half handle on one end. Earl was trying to turn that crank by hand when it suddenly kicked back and the handle wound round the other way. The handle creased Earl on the side of the head and stripped off all his clothes.

Earl was knocked unconscious for several minutes. When he came to, he got up and tried to work, but he was obviously sick and weak and his father sent him on home early.

But still that donkey wasn't working the way it should. Alfred thought the problem was with the friction, so he climbed up on the frame to throw some sand in so it would catch a grip.

One of the cogs caught the cuff of his pants and yanked the pant leg and Alfred's foot into the narrow space between the cam and the frame, crushing the foot.

Fred and the others got the donkey stopped right away, but the mashed foot was crammed in the machinery. It took a while for them to dismantle that part of the donkey to free his leg.

Alfred spent a long time in deep pain at home, hoping the foot would heal. He remembered the episode years before when he'd spent a miserable time in the old hospital in Hillsboro with pneumonia. There was so much draft coming in around the window casings that Hildur brought in blankets from home and nailed them up to stop the breeze. Alfred eventually toughed it through, but the man with pneumonia in the other bed in Alfred's hospital room wasn't quite so lucky. He died.

Hildur nursed Alfred this time, too, but gangrene set in. To save his life, the doctors had to amputate his leg just below his knee.

Fred, barely into his twenties, and Earl, in his late teens, took over the logging operation while Alfred recuperated. He was in pain, and he was angry and grieving and depressed. It was more than a year before he could accept the loss of his leg. But once he did, he got a prothesis and he tramped around the woods again, though he hated to admit how well his sons were running the Nordgren Timber Corporation without him.

Of course, Alfred's sons hadn't told him everything that happened while he was still in the hospital. Fred and Earl had driven the family's Chevrolet coach up into the woods. It was raining some, but together they felled enough trees for a load. One stretch of their logging road was so steep it was really dangerous to try to haul the load out after a rain.

Earl drove the Chevy coach and Fred followed him with the loaded log truck. They stopped at the top of that 30% grade and looked down, knowing they needed the money from this load to help with the medical bills. But they also knew that much of their capital was in the truck and they couldn't afford to wreck it.

"I'll go down first in the coach," Earl suggested, "to see if the truck can make it. Wait here and I'll come back up and tell you if it's all right." Fred nodded, and Earl started down in the Chevy.

It was a wild ride. More than once Earl wanted to steer into the high bank to keep from slipping off the other edge of the road into the canyon, but somehow he made it to the bottom and stopped on the flat. He sat for a moment, blowing through pursed lips. Then he drove the coach in a U-turn and started back up to tell Fred it was too dangerous to try it with the truck.

About half way up he saw the truck starting over the ridge. Fred hadn't waited. He was coming down anyway.

Earl parked the coach at the side of the road out of the way and climbed up the hill a way to watch Fred's progress. The truck was coming down with the wheels turned hard into the bank. The bank was high enough that the truck would only go so far up it and then roll back again in an arc onto the road.

But the road was slippery enough that the truck slid forward some before starting the next arc into the bank. So it was making some forward progress. It was a slick trick, and Earl cupped his hands to yell up to Fred that that was great driving.

Only Fred wasn't driving.

By now Earl could see that Fred had long since abandoned the truck and was running like crazy in the mud behind it trying to catch up.

Earl calculated the truck's driverless sliding path, and knew he was in danger himself. He moved. As he glanced back over his shoulder he could see that the truck was going to smash right into the Chevy coach and both would go over the side into the canyon.

It hit, sending the coach flying over the steep side and out of sight down the canyon wall. But the collision changed the truck's path enough that it didn't go over the side with the Chevy.

And then Earl saw that the window on the driver's side was open. "If I can get in there, maybe I can steer it enough so it won't crash into anything until I can get it stopped at the bottom," he thought. But his reactions were faster than thought, and much faster than the telling.

Earl raced along beside the run-away truck, knowing he couldn't wrestle the door open, so he dived in through that open window and hauled on the wheel to keep them from going over the edge. By that time the slope was leveling off a little. The truck wasn't responding to the brakes, but Earl saw that the airbrake had not been set. In desperation, but with little hope, he yanked on the handle. And the truck slowed and stopped. Safe. The load still intact.

Fred came hurtling down to them, panting. Safe, too.

The two brothers hiked back up and looked over the edge into the canyon. Even the Chevrolet coach had caught on a tree on the way down. The top was smashed in, but resurrecting the rest of the body looked possible, if not probable.

Later when they winched the coach back up onto the road, they were amazed to find that it still ran. At home, they worked with acetylene torches to cut off the top, and the coach rode pretty well.

Later, when Alfred asked the boys about what they'd done to the coach, they told him that it hauled more with the top off like that. But that's all of that story they ever did let their father know.

Hildur Thorin Nordgren

Alfred with Alfhild, Hildur holding baby Manfred

Alfhild, Fred, Earl and Tygue growing up

Carpenter Creek school kids, mostly Persons, Browns, and Nordgrens. (Back: Ted Brown, far L; Alfhild, 2nd from R. Front: Earl, 2nd from L; Maxine Hoover [Schaefer], 4th from L; Fred, far R.)
Below L. Earl on Pat; R. Joe Zberg, Fred, Earl

Earl at 68 years old on top of Mt. Hood